David Douglas. Crayon drawing attributed to Sir Daniel McNee, R.S.A.
Courtesy Linnean Society

DOUGLAS

of the

FORESTS

The North American Journals
of David Douglas

JOHN DAVIES

PAUL HARRIS PUBLISHING

EDINBURGH

First published in Great Britain
1979
by Paul Harris Publishing
25 London Street
Edinburgh

British Library Cataloguing in Publication Data:
    Douglas, David
        Douglas of the forests.
        1. Botany — North America — History — 19th Century
        2. Plant collectors — North America
        3. Scientific expeditions — North America — History
           — 19th century
        I. Davis, John
        917'.04'4   QK110
        ISBN 0-904505-37-5

The publishers acknowledge the financial assistance of the Scottish Arts
Council in the publication of this volume

Printed in Great Britain by The Shetland Times Ltd.
Lerwick, Shetland

CONTENTS

ILLUSTRATIONS

Plates between pages 88 - 89

FOR ANDREW AND ZOE

PREFACE

I came across Douglas' Journals some twenty years ago in the comprehensive book of Douglas' papers published under the auspices of the Royal Horticultural Society in 1914. Although full of good material, the book had been poorly edited and I found it very difficult to read. It struck me that the best of it deserved a bigger public and I set about the task aided and encouraged by many friends and acquaintances.

I should particularly like to thank the Librarians of the Royal Horticultural Society, the Linnean Society and the Royal Commonwealth Society for their patience and kindness. Also the correspondents and helpers in Canada whom I have pestered for photographs and help: Professor Rheumer of Simon Fraser University, Raymond M. Patterson of Victoria, British Columbia, the archivists and librarians of British Columbia, Hudson's Bay Company, the Public Archives of Canada and the Glenbow-Alberta Institute. Also the staff at the Royal Scottish Museum and Dr Ian Hedge of the Royal Botanic Gardens in Edinburgh. Alan Mitchell and Tony Anderson of the Forestry Commission have given me valuable advice and it was kind of His Grace The Duke of Buccleuch to allow me to photograph the famous Drumlanrig Douglas Fir.

Finally I am indebted to Mrs Avril Crosbie of Dumfries, and to my wife for typing my often untidy manuscript.

JOHN DAVIES.
August 1978.

Introduction

The Scot has always believed in the inherent genius of his race and held a special affection for men who have risen to fame from humble origins to prove the point.

The wayward genius of Burns is preferred to the massive achievement of middle class Walter Scott, partly because of this special affection for the underdog. Livingstone, Telford and Alexander Murray of Dunkitterick are all examples of the phenomenon.

David Douglas, the stone-mason's son who left school at eleven and who established himself in his short life as one of the greatest exploring botanists of all time, belongs to this celebrated band. His journals, which tell of his adventures in Western Canada and the United States 150 years ago, have been out of print for many years, and, re-examined, may allow us to appreciate more fully the importance of the man and his work.

David Douglas was born at Scone, near Perth on July 25th 1799, the second son of John Douglas, a stone-mason, and his wife Jean Drummond. They had six children in all, three boys and three girls, and it is recorded that John Douglas was an able man, remarkably well informed for his station in life.

David was only three years old when he first went to dame school and he seems to have been a normal high-spirited child. He soon went on to the parish school at Kinnoull six miles away where he displayed a certain independence of character, much preferring the open countryside, fishing and bird collecting to the classroom. He could read at an early age and recalled later that his favourite books were *Robinson Crusoe* and *Sinbad the Sailor*. He also kept owls and hawks and his general interest in natural history induced his father to apprentice him when he was eleven years old to William Beattie, the head gardener at Scone Palace, the seat of the Earl of Mansfield.

Here he was very happy and picked up a great deal of knowledge from both Mr Beattie and his senior assistant — a Mr McGillivray, who had some basic training in botany. He read a lot, built up a small collection of botanical specimens and was fortunate in getting to know the Brown brothers, who ran the

Perth nursery and who went on annual botanizing trips into the Highlands. There is no doubt that their accounts of these trips and their adventures influenced the boy deeply, although he was to all intents and purposes being trained as an ordinary gardener capable of working in a nursery, with shrubs and trees, and in the kitchen garden or flower garden.

When he had completed his apprenticeship he attended a private school in Perth in order to improve his scientific and mathematical knowledge. He was keen to get ahead.

His desire for self improvement was marked by his taking himself to a private school in Perth once his apprenticeship was over, in order to increase his knowledge of mathematics and science.

William Beattie was fond and proud of David and recommended him to Alexander Stewart, head gardener to Sir Robert Preston of Valleyfield, near Culross in Fife. The great house of Valleyfield and its extensive glass houses have long since disappeared, but in 1818 when David went there the collection of exotic plants was possibly the finest in Scotland.

Stewart warmed to the youth's enthusiasm and he was soon given access to Sir Robert's botanical library and made foreman. In April 1820 he applied for and obtained a post in the Botanic Gardens in Glasgow and quickly won golden opinions from Stewart Murray, the Curator.

In April 1820 another event took place in Glasgow which was of the utmost significance to Douglas' career. William James Hooker took up the Chair of Botany at Glasgow University.

Hooker was born in 1785 and from his early years had been, like Douglas, a keen naturalist. At the age of twenty-one he devoted his prodigious energies to the study of botany and travelled widely in Europe. In 1812 he was made an F.R.S. and during his life he wrote at least thirty works of major importance and literally hundreds of scientific papers. Today he is chiefly remembered for the quite extraordinary way in which he expanded Kew Gardens. In 1841 when he was appointed Director, Kew was a modest eleven acres with ten old fashioned hot houses and conservatories. At his death in 1864 there were 75 acres of botanic gardens, 270 acres of arboretum and pleasure ground and 25 modern hothouses including the massive palm house and temperate house. He is numbered amongst the great scientists of the 19th century.

But Hooker was much more than a great scientist. He was a warm, kindly, family man with a gift for teaching. Charles

Darwin, writing to Hooker's son after his death, remarked on his "remarkably cordial, courteous and frank bearing". He was loved by his pupils and carried on a huge correspondence with them in later years.

In a short time he reorganised the Glasgow Botanic Gardens, and he became one of the most popular lecturers in the University, and many outsiders crowded in to hear him speak. He organised botanical expeditions into the Highlands, and pretty vigorous these were, sometimes covering upwards of thirty miles a day.

David Douglas, who was a glutton for work and exceptionally strong and keen, soon came to the attention of the young Professor. Hooker wrote a memoir of David Douglas which was published in the Companion to the Botanical Magazine in 1836, two years after Douglas' death. If we remind ourselves that this is a brilliant and famous Professor writing about an under gardener we learn something about both of them. Of their contact in Glasgow, Hooker wrote — "Whilst in this situation he was a diligent attendant at the botanical lectures given by the Professor of Botany in the hall of the garden and was his favourite companion in some distant excursions to the Highlands and Islands of Scotland where his great activity, undaunted courage, singular abstemiousness and energetic zeal, at once pointed him out as an individual eminently calculated to do himself credit as a scientific traveller."

David Douglas learned a tremendous amount of practical botany from this splendid man but the real break came in 1823 when Hooker recommended him to Joseph Sabine, at that time Honorary Secretary of the Horticultural Society, to act as a plant collector in China. Douglas left for London but continued to correspond with Hooker and his family for the rest of his life.

An idea of the special relationship may be gained from two letters written to the Hookers from David Douglas. The first, dated March 24th, 1826, from the Great Falls of the Columbia —

"Dear Sir,
From Dr. Scouler you must have obtained a good description of North West America, and be made acquainted with many of its treasures. He left me in fine spirits and when we were together, not a day passed in which you were not spoken of. His departure I much regret; we had always been friends and here our friendship increased. When botanizing along the shores of the Columbia River, and in the adjoining woods, we would sometimes sit down and rest our limbs and then the

conversation often turned on Glasgow and Ben Lomond. If a favourite moss caught his eye, it was eagerly grasped and transferred to the vasculum and the remark was pretty sure to follow — 'How much would Dr. Hooker like to be with us.' I felt very lonely after Dr Scouler had sailed . . . "

The second, dated October 24th, 1832, also comes from the Columbia and contains a delightful postscript to Hooker's elder son William, then aged 16, who later took up medicine and became a promising young doctor. He died of yellow fever in Jamaica in 1840.

"Your kind letter dated just two years ago gives me great satisfaction, as containing accounts of the health and prosperity of yourself, brothers, sisters and parents . . . Your description of the late excursions to Ben Lomond and Killin delights me highly. I only wish I could have been of the party, whether to fish, shoot or botanize. By this time I trust you are almost another Isaac Walton, whose book you should study diligently, if you ever would become a worthy brother of the angle. In California I had fine sport, both at fishing and hunting; the former principally in sea-fish, as those of the river are few and small. This mighty stream (the Columbia) is incomparably the noblest in the world for salmon, trout and sturgeon, whether for quality or abundance. But in the Sandwich Islands, my dear boy, the natives *domesticate* their fish! They catch in the sea, when about two inches long, two kinds of mullet, the grey and the white, with another fish of great delicacy, called in their tongue 'Ava' and remove them to larger ponds of brackish and partly salt water, where they are suffered to remain a few weeks, and ultimately deposited in tanks of fresh water, where they grow exceedingly large and fine, and are taken out for use at the pleasure of the owner. Thus you see these fellows are no despicable fishers.

You may tell your little brother (who wondered that I could bear to go to sea, as there were cockroaches in all ships) that I now feel a mortal antipathy, even more than he, if possible, to these insects; for having made a great number of observations in the Sandwich Islands, the vile cockroaches ate up all the paper, and as there was a little oil on my shoes, very nearly demolished them too!

I have never seen the Aurora Borealis, about which you enquire, particularly splendid, except occasionally near Hudson's Bay; but as I hope shortly to go so far to the North as to see phenomenon in all its magnificence, you shall perhaps hear of it in my next letter.

I trust we may yet have a fine jaunt in the Highlands together, perhaps in the summer of 1835."

(The "little brother" who disliked cockroaches became Sir Joseph

Hooker, who succeeded his father at Kew and consolidated his work there. A very close friend of Charles Darwin, in his early years he explored from the Antarctic to the Himalayas. He inherited his father's prodigious energy and gift for friendship).

Joseph Sabine also needs more than passing mention, for he too had a considerable influence on Douglas. A barrister by profession, he became Honorary Secretary of the Horticultural Society in 1810, some nine years after its foundation, and quickly reorganised its ill-managed accounts and steadily expanded its activities and influence. He was a noted ornithologist and after a time virtually ran the Horticultural Society single handed, but overreached himself and in 1830 was forced to resign. This resignation has considerable implications for Douglas.

When David Douglas reached London he learned that the China expedition had been cancelled because of a breakdown in diplomatic relations. However, in June he was sent to the eastern United States instead, with a remit to collect fruit trees and any unusual botanical specimens. He managed to get up into Canada and brought back some interesting new plants, besides the fruit trees. He returned to London on January 10th 1824.

During this trip he kept a daily journal, which was published, along with the rest of the Douglas papers, in 1914. Many of the entries are of limited interest and space does not permit their inclusion in this book.

The expedition had clearly been a considerable success and a new mission under the joint auspices of the Horticultural Society and the Hudson's Bay Company was mounted in the summer of 1824.

The new expedition was a much more ambitious affair, for he was to explore the states of Washington and Oregon, from a base on the Columbia river.

In Professor Hooker's words:
"We now come to the most interesting period of Mr Douglas' life, when he was about to undertake a long voyage and to explore remote regions untrodden by the foot of any naturalist. In these situations, far indeed from the abodes of any civilised society, frequently with no other companion than a faithful dog, or a wild Indian as a guide, we should have known little or nothing of his adventures, were it not for a Journal which he kept with great care . . ."

David Douglas prepared for this trip in his own way. He worked up to eighteen hours a day improving his scientific know-

ledge. Hooker's faith in the erstwhile under-gardener had not
been misplaced.

On Monday 26th July, 1824 he set sail from Gravesend in
the Hudson's Bay Company's ship *William and Ann,* bound for
the Columbia River.

The ship called at Madeira, Rio de Janeiro, Juan Fernandez,
the Galapagos Islands and arrived at the mouth of the Columbia
on April 17th, 1825, and he recorded enthusiastically all he had
seen on the long and wearisome voyage.

The country was enormous and very largely unexplored.
Vancouver had penetrated the lower reaches of the Columbia
in 1792, and Lewis and Clarke had crossed the Rockies and
wintered miserably at its mouth in 1805. In 1793, further north,
Alexander MacKenzie had burst through across the continent to
the Pacific. In 1808 Simon Fraser descended the great river that
bears his name. In 1811 Thompson of the North West Company
had discovered the Athabasca pass, and in the same year J. J.
Astor had established a trading post at Astoria close to the site
of Lewis and Clarke's camp. The whole area from the Yukon
down to the California border owed allegiance to no one and
was claimed by both the United States and by Britain, for the
agreed line of the 49th parallel did not run west of the watershed of the Rockies. Indeed it was not until 1846 that Oregon
and Washington became part of the United States, and British
Columbia part of Canada.

In 1821 the two rival fur companies, the Hudson's Bay and
the North West merged and during Douglas' time British influence was paramount in the area; indeed Astoria had been
bought over and rechristened Fort George. In 1825, when Douglas landed there were probably no more than 400 white men
in the whole huge territory and the Indians held the whip hand.
On the coast they had been introduced to alcohol by the sea
otter fur traders, but inland European diseases and drink had
not yet taken their dreadful toll and the entire success of the
H.B.C. depended on Indian goodwill. Many marriages took place
and half-breed children were very numerous around the trading
posts.

Douglas was made welcome by Chief Factor McLoughlin
and soon began exploring up the Columbia and some of its
tributaries.

By the end of 1825 he reckoned he had covered 2,105 miles,
had learned to live rough and had established good relations with
traders and hunters, and had made several useful friends amongst

the Indians. Most of the white men were Scots or French Canadians and many of their names are recorded in the mountains and rivers of the region. Douglas' accounts of his meeting with these men are of considerable historical value.

In 1826, now thoroughly established, he reckoned that he covered some 3,932 miles: Spokane, Kettle Falls, The Snake River, Okanagan, south down Willamette River to the Umpqua. The Pacific coast and the Blue Mountains were all extensively explored. He made some wonderful finds and the description of the discovery of the Sugar Pine *(Pinus lambertiana)* is surely a classic adventure story and gives a remarkable acount of the perils that were inherent in the great wilderness.

The early part of 1827 was spent tidying up and sending specimens home. Then on March 20th he set off along the old fur traders' route up the Columbia and across the Rockies to Hudson's Bay. It was a journey marked by hardship and danger and is recorded with wonderful freshness and clarity. Finally at sunrise on Tuesday August 28th — . . . "I had the pleasing scene of beholding York Factory two miles distant, the sun glittering on the roofs of the house (being covered with tin) and in the bay riding at anchor the Company's ship from England." Journey's end.

He sailed from Hudson's Bay on September 15th and arrived at Portsmouth on October 11th, "Having", as he put it later, "enjoyed a most gratifying trip."

Douglas brought back with him an even greater number of plants and seeds than he had sent home and was accorded a warm welcome.

He was made a Fellow of the Linnean, Geological and Zoological Societies, without payment of any fees, and in 1828 Dr Lindley dedicated the primrose genus *Douglasia* in his honour.

Sadly the fame and lionisation of the scientific world of London soon became too much for him and in his 1836 Memoir, Hooker wrote:

"Qualified as Mr Douglas undoubtedly was for a traveller, and happy as he unquestionably found himself in surveying the wonders of nature at its grandest scale, in conciliating the friendship (a faculty he eminently possessed) of the untutored Indians, and in collecting the productions of the new countries he explored; it was quite otherwise with him during his stay in his native land. It was, no doubt, gratifying to be welcomed by his former acquaintances, after so perilous yet so successful a journey, and to be flattered and caressed by new ones; and this

B

was perhaps the amount of his pleasures, which were succeeded by many, and, to his sensitive mind, grievous disappointments. As some further compensation for this meritorious services, the Council of the Horticultural Society agreed to grant him the profits which might accrue from the publication of the Journal of his Travels, in the preparation of which for the press, he was offered the assistance of Mr Sabine and Dr Lindley: and Mr Murray of Albemarle Street was consulted on the subject. But this proffered kindness was rejected by Mr Douglas, and he had thoughts of preparing the Journal entirely himself. He was, however, but little suited for the undertaking, and accordingly, although he laboured at it during the time he remained in England, we regret to say, he never completed it. His temper became more sensitive than ever and himself restless and dissatisfied; so that his best friends could not but wish, as he himself did, that he were again occupied in the honourable task of exploring North West America."

There is ample evidence for Douglas' ill temper in the archives of the Royal Horticultural Society. I came upon a hitherto unpublished account of a bitter row between Douglas, Lindley and Goode during which Dr Hooker reprimanded his protégé. The account, dated 16th April, 1829, concludes with "Mr Douglas appeared quite ashamed of his conduct".

Hooker however still felt very warmly towards Douglas and it was doubtless due to his influence that a little later that year the Hudson's Bay Company made a "most liberal offer of assistance" and aided by a further grant from the Horticultural Society and financial help from the Colonial Office a second expedition to the Columbia was mounted. The Colonial Office support was arranged through Capt. Sabine, Joseph's brother, who in a dark hour befriended him and gave him a good deal of useful instruction in surveying.

Douglas wrote to Hooker on August 6th, 1829:

......"I am sure you will be glad to know that my anticipated journey has been laid before the Council, and approved of; so that I go, God willing, on the 15th of September, by the Hudson Bay Company's Ship Eagle. My plans must be a separate communication, but just let me say, that my principal objects are to make known the vegetable treasures of the interior of California, from the northern boundaries of Mexico, near the head of the Gulf. The botanical productions of Rio Colorado and other streams, totally unknown in Europe, will, I trust, ere many years be as familiar as those of the Columbia. The Government provides me with every instrument which

Captain Sabine, as Secretary of the Royal Society thinks may be of use. These consist of sextants, chronometers, barometers, thermometers, hygrometers, compasses of all sorts, instruments for magnetic intensity, dip of the magnetic needle, all of which can be used with such accuracy, as will render my journey, as I trust, not the journey of a common place tourist.

"I am not quite certain, but that when I have completed my expedition on the Continent of America, I may cross to the opposite shore, and return in a southerly line, near the Russian frontier with China. What a glorious prospect! Thus not only the plants, but a series of observations may be produced, the work of the same individual on both Continents, with the same instruments, under similar circumstances, and in corresponding latitudes! I hope I do not indulge my hopes too far. I shall try to set a hundred pairs of feet, and as many hands to work for me, and shall make them grub up and bring me all they can find. People tell me that Siberia is like a rat-trap, which there is no difficulty in entering, but from which it is not so easy to find egress. I mean at least to put this saying to the test. And I hope that those who know me know also that trifles will not stop me. I am glad to learn you are coming to England, before I go, that I may see you once more. I shall be greatly obliged if you would purchase for me a Bible, in 2 vols. 8vo., with a good bold legible type, and notes of reference, or more properly speaking, marginal notes. I cannot see to read small type, and have been unable to find such an one in London, but I know there is a Scotch edition of the kind which I describe."

The ambition to walk back to Europe across Siberia was to preoccupy him a good deal during this last period of his life.

When the Journals were published in 1914, the editors added a memoir which describes this final expedition:

"The President and Council were so gratified by the result of Douglas' expedition that they persuaded him to return, which he did in October 1829 with the intention of undertaking a wider exploration in the same general districts as before, but the unsettled state of the country and the tribal wars going on among the natives made this impossible. He therefore transferred his attention from the Columbia river to California and landed in San Francisco in 1831, whence he journed to Monterey, where he was well received by the monks and afforded every facility in their power for exploring the neighbouring country. He remained there during the summer of 1831, intending to return to the Columbia river in the autumn; but being unable to find any ship or other means of transport he was compelled to spend another season in California making various excursions into the interior, and finally in August 1831 he sailed to the Sandwich Islands (Hawaii), from whence he despatched his Californian collections of seeds and plants and later on returned to the Columbia River. From this collection more species were raised in the Society's gardens.

On his way from the Sandwich Islands to the Columbia River he received intelligence that his personal friend, Mr Sabine, had resigned the secretaryship of the Society, and through some misconception of the cause Douglas also resigned his appointment of Collector. For more than a year however he continued travelling in North Western America."

The Fort Vancouver Douglas returned to in 1833 had changed greatly over the past eight years. Trading had given way to settlement and the beginnings of forest clearance — the first American pioneers were trickling through. Smallpox and measles had decimated some of the Indian settlements. In spite of failing eyesight Douglas decided to have one last attempt to work his way north to Alaska and with Russian help, walk back to Europe across Siberia.

He ascended the Columbia, cut across to the Fraser River and got as far north as Fort St. James on the shores of Stuart Lake. But he could get no further. The Indians were in a state of insurrection and he could get little assistance from the trading post staff. Reluctantly he turned back and below Fort George disaster struck, when he was wrecked and nearly drowned in the rapids. He lost, and the scientific world lost too, his precious Journal for the last years. His specimens and equipment were ruined. Sadly, and broken, he returned to the Columbia.

A little later he set off again to the Sandwich Islands on the Hudson's Bay trading ship. Arriving there on January 2nd, 1834, he made arrangements to climb Mouna Loa, a volcano some 13,800 ft. high. He reached the summit on January 29th, and his letter to his brother John, who was Clerk of Works Drumlanrig Castle in Dumfriesshire was the last correspondence he sent home.

On July 12th, 1834, he was traversing one of the Islands when he fell into a pit dug for wild cattle and was tramped and gored to death by a trapped bull.

There are in fact several theories about his death, and even as recently as 1970 fresh evidence was brought forward to suggest murder. I think that it is highly unlikely, and that Douglas, insatiably curious and short sighted, went too near to the edge of the pit, and slipped in.

He was buried in Hawaii on August 4th, 1834, after a post mortem.

About 70 years ago a chest containing David Douglas' papers was rediscovered amongst the archives of the Royal Horticultural Society, and Mr Wilks, the Secretary, was requested to edit and publish them on behalf of the Society. It proved to be a difficult task, and in his Preface to the resulting book, which was finally published in 1914, Wilks wrote:

"Many causes have contributed to the delay which has taken place in the publication of the Diaries of David Douglas's journeys on behalf of the Royal Horticultural Society:

1. The handwriting is nowhere easy to read, and in places is most difficult, occasionally almost if not quite impossible.

2. In the course of nearly one hundred years the ink has faded and become in places very hard to decipher.

3. After the Diary of his journey in North Western America had been prepared for the press and set up in type, a second manuscript was discovered which was at first sight taken to be a duplicate, but which on closer examination was found to contain a great deal of additional information. It had therefore to be compared word for word with the Diary and the additions inserted in their proper places.

4. All the botanical names mentioned have been very carefully looked up by Mr M. R. Hutchinson and the name given to the plant in Index Kewensis or by other later authority is quoted at the bottom of the page with a reference to the author responsible for it.

5. Last, but by no means least, the work was entrusted to the Secretary of the Society and to Mr Hutchinson, the Librarian, each of whom had already more work in hand than he could conveniently compass, and only occasionally spare time could be given to a work which had already waited nearly one hundred years for publication before it was committed to their charge."

The Royal Horticultural Society has most generously allowed me free access to the Douglas papers and I can testify to their unsatisfactory condition. One can see the pencilled editings of 70 years ago and the confusion about which of the two versions is the original one. The original version is mercifully written on paper with an 1824 watermark, and the revised edition (which Douglas was no doubt preparing for publication by John Murray) is written on paper with an 1828 watermark. I have only drawn on the revised edition for part of the sea voyage in 1824. All the North West America material is taken from the original working journal.

The 1914 book is nevertheless very badly arranged and almost as difficult to read as a Victorian bible. Naturally it is compiled by botanists or horticulturalists for their own kind, but I am sure that if it had been better produced the Journal would have been far more widely known and become perhaps a minor classic of exploration. The publication date, too, August 18th, 1914, cannot have helped, for the attention of most Britons and British scientists would have been concentrated on more pressing affairs on the continent of Europe. From the Journals and contemporary accounts a picture of David Douglas emerges that is, to say the least, at variance with the fictional portrait depicted in William Norwood's *Traveller in a Vanished Landscape.*

He was clearly an exceptionally tough, resolute and brave man. One detects a certain obstinacy — in the Scots idiom he could be a thrawn devil — but he had deep loyalties and to those that he admired, such as Sabine and Hooker, he showed a gentle side to his nature. He remained devoted to his family and was good with children. In sophisticated society he could be rude and churlish, but in the country of the far West he showed a tolerance and understanding of the greatly diverse people he met. He never seems to have doubted his Christian faith, and was, unlike some of his race, entirely unbigoted.

It used to be said of a particularly famous and awkward mountaineer that a Society should be formed to keep him above 20,000 feet, because he became quite delightful in adversity at high elevation! David Douglas was at his best whilst engaged on his travels.

Douglas' achievement as a plant collector is extraordinary, and his discoveries and introductions have done a great deal to alter and enrich the economic pattern of our countryside, particularly with forest trees. His introductions into Britain included Sitka spruce, the most widely planted of all trees nowadays, and what is more remarkable, he foresaw with his gardener's eye, its economic importance — "it possesses one great advantage by growing to a large size on the Northern declivities of the mountain, in apparently poor thin soils. This unquestionably has great claims on our consideration as it would thrive in such places in Britain where even Scots pine finds no shelter. It would become a large and useful tree."

In the early 1930s the Forestry Commission recognised the

vital role of Sitka spruce in British silviculture and Douglas'
vision was realised. And then there was Douglas fir itself, the
Noble Silver fir, Lodgepole pine and the Grand Silver fir, all
of which are widely planted in Britain. He also foresaw the
importance of and introduced the Monterey pine *(Pinus radiata)*
which has been used to form fast growing plantations in many
countries of the southern hemisphere.

Douglas holds a very special, indeed unique place in the
affections of professional foresters in the temperate zone. He is
also something of a folk hero among the timber men of the
North Western States to this day.

Then there was the purely botanical contribution, the man's
genius for finding new plants. He was a first-rate field naturalist
and had learned under Hooker how to recognise subtle changes
in habitat which might yield undiscovered treasure. An appendix
in the first 1914 publication of Douglas' Journals lists 254 plants
introduced by him into Britain. Many of our most popular
garden plants and shrubs were discovered by Douglas or bred
from his discoveries. Some, such as the red flowering currant,
are now accepted as native.

But it was not only botany that absorbed him. He was, as
I said earlier, a field naturalist of the first rank, and an
excellent ornithologist. The accounts of his long journeys by
ship reflect this interest, and the little boy who kept owls grieved
when he lost his tame Calumet eagle after taking it 2,000
miles across Canada. Of his eight published papers, one is on
the Californian vulture and one on two hitherto undescribed
species of North American mammal. This versatility was typical
of the scientists of his age — Banks — Hooker — Darwin —
amateur or professional. They took all nature as their kingdom
and brought to it an enthusiasm and sense of wonder that we
have lost in this deeply specialised one. You would need forty
modern scientists to bring back as many new species from an
unexplored country as one Darwin or Hooker. And I doubt if
any of them would have the breadth of vision to advance a
great new truth.

I have edited the Journals fairly severely — they served
Douglas as a aide-memoire as well as a record designed for
publication. He was not a brilliant writer, but there are some
passages of genuine literary merit. Although I have had to cut
out most of his botanical notes, I have only altered his text
in order to clarify the reader's understanding with regard to
species or place names. For example his *Exocoetus rolans* I

have given as "flying fish", his *Diomedea exulans* is "Wandering albatross", his Multnomah which is the Willamette, and so on. With regard to coniferous trees I have had a major difficulty. Douglas called nearly all his conifers Pines, whereas in fact they consisted of several genera. Where I am certain what he saw I have given the modern name, where uncertain, his name.

The Journals have a considerable value to students of the history of the North West. In the 1820s there were very few white men living in Western Canada, Oregon and Washington, but they have left their names indelibly on the place names of a huge territory. Douglas met many of them, and travelled with many of them. His records deserve to be freely available. The Hooker correspondence too throws more light on the personality of that most lovable of men.

However, to me their greatest interest lies in the account of the travels of a most observant man through a primitive wilderness — of journeys by canoe or on horseback — of icy rivers and freezing wind — and long portages across terrible terrain. He was often in danger from Indians, or bears or starvation. He was content to travel on his own for long periods, but he was also a social animal and a good companion. Enormously tough and single minded, he never lost a sense of innocent wonder at the riches of Creation.

I first stumbled on the 1914 book some 20 years ago. It was, as I have said, very badly put together, and rather like the Bible, difficult to read intelligently. Nevertheless, it has given me a great deal of pleasure and I hope that this rearranged edition may be widely read and give to others rather more easily some of the enjoyment that I have received. And it was all such a short time ago.

1

The Journey Out

David Douglas returned to London from his first plant collecting expedition to the eastern United States and Upper Canada on January 9th, 1824. He had done well, and far exceeded the expectations of the Fellows of the Horticultural Society, having brought back a large number of botanical specimens and also several interesting species of birds. It was also clear that he had shown considerable tact.

For some time British landowners, gardeners and botanists had been wanting to obtain specimens and seed from the Pacific North West. They had been excited by the material that Nelson and Anderson had sent back from Captain Cook's expedition in 1778, and by Archibald Menzies' specimens — the surgeon naturalist who accompanied Captain George Vancouver on his ship *Discovery* on the surveys of 1792-1794.

Joseph Sabine, the Secretary of the Horticultural Society, approached the Hudson's Bay Company in the spring of 1824 to afford facilities for a botanist to collect plants in its huge territory. The Company readily agreed and Douglas was appointed and given a passage out on the annual voyage to the North West. He spent the four months before he sailed working desperately hard to improve his scientific knowledge, and his knowledge of the history and geography of the region.

He found to his delight when he joined his ship at Gravesend that an old friend and fellow pupil, John Scouler of Glasgow, had on Dr Hooker's advice, been appointed surgeon/naturalist for the voyage. The route ran from Gravesend to Madeira, Rio de Janeiro, Cape Horn, Juan Fernandez and the Galapagos Islands to the Columbia River. The Journal gives a fair idea of the excitement the two young naturalists experienced collecting sea birds and other specimens.

Saturday July 24th 1824   After several weeks' preparation for a voyage to the Columbia river on the west coast of North America, on the afternoon of Saturday parted with J. Sabine,

Esq., and all other friends. In the evening wrote a letter to my father, to Dr Hooker, and Mr Murray of Glasgow.

July 25th Left London at half-past eight o'clock in the morning from the Spread Eagle office, Piccadilly, by the Times coach in company with my brother for Gravesend.

The morning was very pleasant, cloudy and calm. Passed some fields of rye, cut down; wheat, oats, and barley nearly ready for the sickle. At Gravesend I met Mr John Scouler of Glasgow, who was going on the same voyage to officiate in the capacity of surgeon. This was to me news of the most welcome kind, being previously acquainted with each other and on the strictest terms of friendship. At twelve o'clock went on board the Hudson's Bay Company's ship *William and Ann,* Captain Hanwell, bound for the Columbia river, north-west coast of America, and came on shore again at two o'clock. At 4 o'clock in the afternoon saw my brother in the steamboat for London, who was affected at parting with me, and returned to the ship.

Monday 26th In the morning, employed stowing away all my baggage etc. Went under weigh at four o'clock p.m., having a fine breeze with rain; thermometer 58°. We made only seven miles and then let down the anchor at darkening.

Tuesday 27th Cold with thick fog; passed the Nore at daylight; at seven the ship struck on the "Shivering Sands" and beat about dreadfully for an hour. Fortunately the wind was moderate with little swell at the time, otherwise our situation must have been perilous. On being rescued from our unhappy situation, it afforded the captain and pilot, as well as all on board, much pleasure to learn that the vessel had sustained little or no injury. I confess it gave me pleasure to be enabled to proceed, as delays in such undertakings are by no means agreeable. The pilot left us off Deal at six o'clock in the evening (thermometer 63°). Did not write as the captain intended to put into Portsmouth, to await the orders of the Company as to his proceeding to sea.

Thursday 29th Passed Dungeness having a very favourable breeze; at noon a perfect calm off the Isle of Wight. The vessel being in much better condition than was anticipated, the captain abandoned the idea of putting in at Portsmouth and would go to sea at daylight. The vessel made only 2 inches of water during

the last twenty-four hours. From the appearance of the sky
it must have been a very warm day on shore. Thermometer at
eight o'clock this morning 57°, at twelve 73° — at eight in
the evening 64°.

Friday 30th   A light air of wind sprang up at midnight, and
before daylight the Portland lights were observed. Middle of
the day calm and warm (thermometer at 70°). At noon fine
wind, which continued during the night.

Saturday July 31st   Morning cool, accompanied with rain; midday warm and dry. Passed the Lizard at eleven o'clock. In the
afternoon a strong wind from the south, with intervening
showers of rain. At twelve o'clock the thermometer was 62°;
at 4 p.m. 57°. In the evening our delightful view of the rocky
shores of Cornwall closed.

Sunday August 1st   During the night strong wind from the
north east with heavy showers of rain; at eight this morning the
sun broke through and I had an observation; 8° west longitude.
Towards afternoon wind moderate, and pleasant in the evening.
Passed a schooner at four o'clock. During the whole of our
progress down Channel only saw about fifty gulls and a few
other sea-birds. Thermometer at five in the evening at 66°.
Our scenery, sky and water, but in these a great variety is seen.
At night when there is a gentle ruffling on the water the
Medusae, Physalae and other zoophytes giving off their phos-
phoric or illuminating particles over a vast expanse of water
produce a very fine effect.

August 3rd   Wind from the north, rather cold (thermometer
62° at twelve o'clock); sun visible. Lat. 45° 32' Long 12° 43'.
A year has elapsed today since I arrived at Staten Island, near
New York. It was warm and pleasant and afforded gratification
after a tedious passage of fifty-seven days from Liverpool. Is
there anything in the world more agreeable to the feelings of a
prisoner than liberty? Saw only two birds, resembling gulls; they
seemed shy.

August 9th   About 4 p.m. had the pleasure to anchor in the
Bay of Funchal. So far as the experience of two days went on
this highly delightful island I was amazingly gratified; visited
the summit of one of the highest mountains in company with

my companion the doctor; collected for our respective herbaria
several interesting though not new plants; visited the vineyards
in the neighbourhood of the town, hospital, church and other
establishments, when we resumed our voyage in the evening
of the 12th towards Rio de Janeiro. From August 12th to 24th
as we approached the Equator, the temperature increased — the
maximum 76°, minimum 59°; the greatest height was 84° in
the shade on the 21st at 3 p.m.

The mornings were particularly pleasant and fine. Near
to the Cape Verde Islands Flying fish were skipping
from wave to wave and sometimes fluttering on board, our vessel
being low in the water, more especially in the evenings; and the
screaming noise of Red billed tropic birds with the never absent
Storm petrels, were the only alleviation from sky and
water. Ten degrees on each side of the Equator the weather was
very variable; calms, thunder and lightning, and sudden gusts
of wind, which made this part of our voyage somewhat tedious.

On September 3rd the south-east trade sprang up, which
carried us to within sight of Cape Frio on the 26th. Towards
evening the ship was surrounded by a vast variety of sea-birds
—*Procellaria*—and for the first time I saw a Wandering albatross.
The magnificent prospect to the harbour of Rio is well known.
One feature in Brazilian scenery strikes the European eye —
that is, the leading one, the palms being larger on the summit
of the highest hills. During my stay I had the pleasure to become
known to William Harrison, Esq., residing at Botafogo, through
whose exertions many beautiful plants have been introduced to
England, and who bestows great pains in procuring subjects in
other departments of natural history illustrative of Brazil. In
company with this gentleman and Mr Henry Harrison, a
relation of his, who afforded me great assistance. I made a short
journey to Tijuca, where I was extremely gratified with the rich
luxuriance of the forest (though seen to disadvantage, being too
early), and was particularly delighted with the varied and end-
less forms of *Orchideae*.

Mr Harrison cultivated with great success about seventy
species of this family of plants by being simply nailed to his
garden wall, having the benefit of the bark or wood on which
they grew still given them for support. He had also an aviary
containing several rare as well as beautiful native birds. I became
known also to John Dickson, Esq., surgeon in the Royal Navy,
who was never more happy than when he had an opportunity
of doing an act of kindness. I left on October 15th this charming

place with much regret, more especially having scarcely been able to put myself in possession of any dried plants from the earliness of the season and from the continued rain. For a few days, until we got clear of the land the weather was changeable, accompanied by rains in the evenings.

A 4 a.m. on Tuesday the 19th a fine breeze sprang up and we bore away for the south, gradually leaving the warm weather. Off the Plate, in lat. 37° S., long 37° W., immense shoals of *Fucus pyriformis* passed the ship, some of which measured 60 feet in length, and a stem at the thickest part 3 inches in diameter.

Saturday November 5th Off the Falkland Islands in Lat. 54½° S. We now began to feel the chilliness of Cape Horn and experience the bad weather of its forbidding, dreary climate. It is only when the wind blows furiously and the ocean is covered with foam like a washing-tub that I could take the Albatross. Diametrically opposite to every account I have read of them, they all say calm. Their voice is like the bleating of goats; on being taken they emit from the mouth an oily matter of different colours, arising no doubt from the great variety of *Physalae, Beroe,* and other zoophytes on which they live.

In all, of the brown ones I caught forty-nine, two of which I preserved; both males; no females came under my notice. Off the Cape a third species made its appearance, white on the belly and under the wings, back greyish, blackish-brown on the upper side of the wings; neck light azure colour; beak, black upper part and point yellow; legs and feet black. Two of this I caught, but only one could be preserved (a male); it is a much larger and stronger bird than the other: when he attempts to take the bait, or even to light near it, they all, seemingly with fear, leave it to him. He was very ferocious and would bite at sticks held out to him; one of the sailors in assisting me to lay hold of him was bitten in the thigh through the trousers — the piece was taken out as if cut with a knife. Their flesh is fishy and rancid. On the same day I caught two petrels of a blackish-white colour, beak and legs partly red; this species on the water is very graceful and by no means very plentiful, their voice is like the chuckling of young ducks; the two now sent home are males.

During the time (ten days) of rounding the Cape the weather was stormy with generally a fine clear sky. The motion of the vessel was great, the waves frequently breaking over it, and no sleep until completely worn out with fatigue. When the wind blows from the south or south-west the cold is insupportable, and yet the thermometer never was lower than 39°, 45° the greatest; there is a piercing rawness in the atmosphere (laying aside being so lately in the tropics and of course more susceptible of cold) quite unknown in the northern hemisphere in similar latitudes. Daylight sixteen hours, sky generally clear azure and beautifully tinted in the evenings just as the sun leaves the horizon.

November 16th We were considered round, and gladly we bade adieu to such inhospitable regions. The weather moderated gradually and we found ourselves navigating more pacific water.

November 17th Caught two albatrosses; the largest weighed 18 lb. and measured 12 feet 4 inches from tip to tip; 4 feet from the point of the beak to the tail. Both were moulting and not worth preserving. All the species when sitting on the water raise the wings like the swan — when eating particularly so — but do not shake them like the hawk tribe. When rising from the water they partly run, partly fly, tipping the water with the point of the wings and feet for several hundred yards before they are clear of it. They cannot rise from the deck of the vessel. Their flight is quick but steady; when fishing they soar with wings in a curved direction.

December 14th Towards noon the Island of Mas-a-Fuera was seen, distant about seven leagues, and appears like a conical black rock. As we drew near the shore it became more like an island. At four in the afternoon of the same day passed within two miles of it; the surge on the beach prevented the commander from landing. On the whole its appearance is barren, although in the valleys there is herbage and some trees on the hills; goats were seen in abundance. Our course was then directed towards the island of Juan Fernandez, about eighty miles distant to the north-east.

It afforded me much gratification to see Juan Fernandez on the morning of the second day. At twelve o'clock a boat was sent in search of fresh water, in which I was permitted to embark; being unsuccessful in some measure, our stay was short and we returned to the ship in a few hours. The following day we went round to the north side to Cumberland Bay, so named by Anson in 1741. The whole island is very mountainous, volcanic, and beautifully covered with wood to the summit of the hills, the tops of which are rarely seen, being enveloped in the clouds. On Friday and Saturday I went on shore and was gratified with my visit. As we approached the shore we were surprised to see a small vessel at anchor, and on the beach a hut with smoke rising from it. As we were about to step out of the boat a man sprang out of the thicket to our astonishment and directed us into a sheltered creek. He gave me the following account of his adventures. His name, William Clark; a sailor; native of Whitechapel, London; came to the coast of Chile five years ago in a Liverpool ship called *Lolland,* and was there discharged. He is now in the employment of the Spaniards, who visit the island for the purpose of killing seals and wild bullocks, which are both numerous. Five of his companions were on the opposite side, in their pursuit, and came to see him once a week; he was left to take care of the little bark and other property. When he saw the boat first he abandoned his hut and fled to the wood thinking us to be pirates. On hearing us speak English he sprang from his place of retreat, and no language can convey the pleasure he seemed to feel. He had been there five weeks and intended to stay five more; he came from Coquimbo, in Chile. His clothing was one pair of blue woollen trousers, a flannel and a cotton shirt, and a hat, but he chose to go bareheaded; he had no coat. The surgeon and I gave him as much as could be spared from our small stock, for which he expressed many thanks. His little hut was made of turf and stones thatched with the straw of a wild oat. In one corner lay a bunch of straw and his blanket; a log of wood to sit on was all the furniture; the only cooking utensil was a common cast-iron pot with a wooden bottom, which he had sunk a few inches in the floor — and placed the fire round the sides! He longed to taste roast beef (having had none for seven years) and one day tried to indulge with a little baked, as he termed it; but in the baking the bottom gave way, as might reasonably be expected; so poor Clark could not effect the new mode of cooking. I told him under his circumstances roasting beef was an easier task than boiling. He is a man of some information; his library amounted to seventeen volumes — Bible and Book of Common Prayer, which he had to keep in a secret place when his Spanish friends were there; an odd volume of 'Tales of My Landlord' and 'Old Mortality', some of voyages,

Cowper's Poems. He had the one by heart addressed to Alexander Selkirk; but what is still more worthy to be noticed, a fine bound copy of Crusoe's adventures, who himself was the latest and most complete edition. Like all other English sailors he had no aversion to rum; I gave him a single dram, which, not being accustomed to before for a great length of time, made him forget his exile. He was like the heroes of Troy; 'fought his battle over again and slew the slain three times'.

Here a few years ago the Spaniards formed a colony; but it is now abandoned, all the houses are destroyed, and the fort, on which were some very large guns. Twenty-six cannons lay on the shore just below. The vestiges of a church are to be seen; on the lintel of the door the following inscription, 'La casa de Dios puerta del dielo y so colocoesta a 24 de Septiembre, 1811' — 'The house of God consecrated 24 September 1811'. Near this a circular oven of brick, seven feet within, marked on it 1741; probably built by Anson during his residence, it is now occupied by a small species of blue pigeon as their cote; in it I found some eggs, but no young ones. This I told Clark he should use. In the old gardens were abundance of three or four different peaches in a half ripe state, very luxuriant; one apple, a quince, and two pears; a quantity of the last three we took for puddings. Abundance of figs in vigorous state of bearing, and vines, one which thrives luxuriantly; it is just in blossom. The only ripe fruit was a sort of strawberry with large fruit of a pale whitish-red, not unpleasant; leaves, stem, and calyx very downy; dried a paper of seeds of this species lest it may prove indigenous to the island or the coast of Chile. The only culinary vegetable was radish, which grows to a large size. I sowed a small portion of vine, pear, and some other fruit seeds which I had of Mr Atkinson and some culinary vegetables, and gave some to Clark to sow on various parts of the island. Saturday afternoon was set apart for fishing; a sort of rock codfish and a smaller fish unknown to me were caught in abundance, both good eating, and after such a length of time on salted food were considered a luxury.

On our quitting the shore Clark presented us with a fine female goat, but not one of Robinson Crusoe's, for it was young. We left him standing on a large stone on the shore on the evening of Saturday, intending to visit him again in the morning. Scarcely had we reached the ship when a strong south-easterly wind set in, which obliged me reluctantly to leave such an interesting speck of the globe and my new acquaintance Clark.

The weather continued unfavourable for making the land again; for three days we were so much driven by its violence that the captain considered it a sacrifice to return. Our course was then directed towards the Islands of Galapagos under the Equator in Long. 80° W. On the morning of Thursday the wind became moderate and we got the south-east trade wind, which we were fortunate enough to carry with us within 1½ degrees of the Equator. Christmas was observed in Lat. 27°S., Long 84° W. We dined on the goat given to us by Clark; were comfortable and happy; in the evening we drank the health of our friends in England.

At noon on Sunday, 9th, Chatham Island was seen; we passed along the east side at 4 p.m. of the same day, fifteen miles from the shore. It is not mountainous and apparently but little herbage on it. On the morning of 10th (Monday) I went on shore on James Island, thirty-seven miles to the west of Chatham Island. It is volcanic, mountainous, and very rugged, with some fine vestiges of volcano craters and vitrified lava; the hills are not high, the highest being about 2000 feet above the level of the sea. The verdure is scanty in comparison with most tropical climates, arising, no doubt, from the scarcity of fresh water, although at the same time some of the trees in the valleys are large, but very little variety few of them were known to me. My stay was three days, two hours on shore each time. Few of its plants were known to me. The birds are very numerous, and some of them pretty, so little acquainted with man's devices that they were readily killed with a stick; a gun was not necessary except to bring them from the rocks or from the tops of the trees. Many of the smaller ones perched on my hat, and when I carried my gun on my shoulder would sit on the muzzle. During my stay I killed forty-five, of nineteen genera, all of which I skinned carefully, and had the mortification to lose them all except one species of Booby; by the almost constant rain of twelve days after leaving the island I could not expose them on deck and no room for them below. Among them were two species of pelican, four of booby, four of hawk (one particularly fine, nearly orange colour), one very curious small pigeon. I was nearly as unfortunate with plants, my collection amounting to 175 specimens, many of them, no doubt, interesting. I was able only to save forty. Never in my life was I so mortified, touching at a place where everything, indeed the most trifling particle, became of interest in England, and to have such a miserable collection to show I have been there.

C

On the island were a species of tortoise, some of them very large, one weighing 400 lb; a lizard, 3 feet long of a bright orange yellow; both good eating. A fine skin of the lizard I lost, and regret it exceedingly, being not described. On the shore are abundance of turtle of good quality, probably the green turtle of the West Indies. No fresh water was found except a small spring flowing from the crevices of one of the craters. The last day on shore it ceased to rain for about an hour; the sun broke through and raised a steam from the ground almost suffocating. My thermometer stood at 96°, not a breath of wind.

On the 12th of February we were in sight of a river in Long. 134° W., but the weather was so boisterous and frightful that it forbade everything like approaching the coast as useless. We were tossed and driven about in this condition for six weeks, winds prevailing from the south-west. Here we experienced the furious hurricanes of North-West America in the fullest extent a thousand times worse than Cape Horn. In this latitude there is an abundance of a small species of *Physalae* of an azure transparent colour, which were frequently washed on the mainyard by the spray breaking over the vessel. Prevailing winds from the south-west and north-west. Many efforts were made during this time to reach the destined port in the short intervals of favourable weather. On Saturday, April 2, Cape Disappointment was seen at noon, distant thirty miles. Sail was shortened to wait a new day for entering.

Sunday 3rd Calm in the morning and cold; a keen easterly breeze carried us within four miles of the River, when another violent storm from the west obliged us again to put to sea.

April 5th We bore in again for the land, being 170 miles at sea, with weather more inviting.

April 7th At daylight on Thursday our course was again directed to the coast, being only 40 miles distant, every person breathing a wish we might be more fortunate than on Sunday. The weather seemingly more steady with a keen north-east wind, such an opportunity was not lost, all sail was set, joy and expectation was on every countenance, all glad to make themselves useful. The Doctor and I kept the soundings. At one o'clock noon we entered the river and passed the sand bank in safety (which is considered dangerous and on which, I learn, many vessels have been injured and some wrecked). At four we came to anchor in Baker's Bay, on the north side of the river.

Several shots of the cannon were immediately fired to announce our arrival to the establishment 7 miles up the river, but were not answered. Thus my long and tedious voyage of 8 months 14 days from England terminated. The joy of viewing land, the hope of in a few days ranging through the long wished-for spot and the pleasure of again resuming my wonted employment may be readily calculated. We spent the evening with great mirth and at an early hour went to sleep, to sleep without noise and motion, the disagreeable attendants of a sea voyage. With truth I may count this one of the happy moments of my life. As might naturally be supposed to enjoy the sight of land, free from the excessive motion and noise of the ship — from all deprived nearly nine months — was to me truly a luxury. The ground on the south side of the river is low, covered thickly with wood, chiefly *Tsuga canadensis*,[1] *Abies balsamea*,[1] and a species which may prove to be *Pseudotsuga menziesii*[1]. The north (Cape Disappointment) is a remarkable promontory, elevation about 700 feet above the sea, covered with wood of the same kind as on the other side.

1. The trees were actually *Tsuga hetecophylla, Abies amabilis* and Douglas fir.

2

First Impressions

Scouler and Douglas landed and found that the old trading post of Fort George (previously Astoria, founded by J. J. Astor for American fur trading) was being abandoned and that a new post, Fort Vancouver, was being established 80 miles upstream, near the outfall of the Willamette River. Chief Factor John McLoughlin M.D., one of the great figures of the North West, was in charge of all this and he made Douglas very welcome.

Basing himself on Fort Vancouver Douglas began to make expeditions up and down the Columbia, at first attaching himself to trading parties of white men, but later, as he grew more confident, just with a Indian guide. By the end of 1885 he had become an expert explorer and had covered over 2,000 miles. He had made some remarkable discoveries, and he clearly possessed an unusual ability to get on with the Indians.

April 8th 1825 Constant heavy rain, cold, thermometer 47°. Saturday the 9th in company with Mr Scouler I went on shore on Cape Disappointment as the ship could not proceed up the river on consequence of heavy rains and thick fogs. On stepping on the shore *Gaulthera shallon* was the first plant I took in my hands. So pleased was I that I could scarcely see anything but it. Mr Menzies correctly observes that it grows under thick pine-forests in great luxuriance and would make a valuable addition to our gardens. It grows most luxuriantly in the margins of woods, particularly near the ocean. *Rubus spectabilis* was also abundant; both these delightful plants in blossom. In the woods were several species of *Vaccinium,* but not yet in blossom, a species of *Tiarella* and *Heuchera* in flower. In a few hours we returned to the ship amply gratified.

Sunday the 10th went again on shore and made a short stay; saw nothing different from that seen yesterday, except some *Gramineae* and *Musci.* On our return to the ship we found a canoe with one Canadian and several Indians with intelligence from the establishment who brought some potatoes, milk, and fresh butter. The potatoes were so much relished that we had some in the evening for tea. The natives viewed us with curiosity and put to us many questions. Some of them have a few words of English and by the assistance of signing make themselves very well understood. The practice of compressing the forehead, of perforating the septum of the nose and ears with shells, bits of copper, beads, or in fact any hardware, gives a stranger a curious idea of their singular habits. They brought dry salmon, fresh sturgeon, game, and some prepared roots with dry berries for sale and soon showed themselves to be a dexterous people at bargaining. On Monday, the 11th, we went up the river to the Company's establishment, distant from the entrance about seven or eight miles. We learned they had nearly abandoned their fort there and had made one seventy miles up the river on the opposite side, to which all persons in their employ were to repair in a few days. I went on shore on Tuesday (12th April, 1825) and was very civilly received by a Mr McKenzie, the other person in authority; he informed me they were about to aoandon the present place for a more commodious situation 90 miles up the river on the north side, also that the chief factor, John Mc-Loughlin, Esq., was up the river at the new establishment, but would be down as soon as he received the news of the ship's arrival. I did not leave the ship until Saturday, but was daily on shore. With respect to the appearance of the country and its fertility my expectations were fully realised. It is very varied, diversified by hills and extensive plains, generally good soil. The greater part of the whole country as far as the eye can reach is closely covered with pine of several species. In forest trees there is no variety or comparison to the Atlantic side, no beech, gleditschia, magnolia, walnut, one oak, one ash, The country to the northward near the ocean is hilly, Point Round or Point Adams of Lewis and Clarke on the south side of the river is low and many places swampy. For the distance of 40 miles as far as Cape Lookout there is a ridge of hills that run in a south-west direction and is so named by Vancouver. The breadth of the river at its mouth is about 5 miles not including Baker's Bay which has a deep bend. The current is very rapid and when the wind blows from the west produces a great agitation. The water on the sandbar breaks from one side to the other so that no channel can be perceived; when in such a state no vessel can attempt to go or come in. Mr Mackenzie made me as comfortable as his circumstances would admit, until he could see the chief factor. My paper being all in the hold, except a very small quantity, and the ship not yet taking out the cargo, I could do but little in the way of collecting.

Saturday 16th April 1825　The chief factor, John McLoughlin, Esq., came down the river from the new establishment, who received me with much kindness. I showed him my instructions and informed him verbally the object of my voyage, and talked over my pursuit. In the most frank and handsome manner he assured me that everything in his power would be done to promote the views of the Society. Since I have all along experienced every attention in his power, horses, canoes and people when they could be spared to accompany on my journeys. Also in every instance had much assistance from those in authority under him, with all the comfort the country affords — circumstances which I am confident that it will give Mr Sabine much pleasure in communicating to the committee of the Hudson Bay Company. The same day I had all my articles sent on shore from the ship. Mr McLoughlin advised me to visit the new establishment as they were shortly to abandon this one on the coast.

Tuesday 19th　In company with him I left the mouth of the river; at 8 o'clock morning in a small boat with one Canadian and six Indians; we made only forty miles, having no wind and a very strong current against us. We slept in the canoe, which we pulled up on the beach. Our supper was a piece of good sturgeon, a basin of tea, and a slice of bread. We had six Indians for paddling the canoe; they sat round the fire the whole night roasting sturgeon, which they do by splitting a branch and placing the meat in it, twisting a bit of rush at the top to prevent it falling out. They ate a fish weighing about 26 or 28 lb. from ten o'clock at night till daylight the following morning. They had young shoots of *Rubus spectabilis,* and water. We started at three o'clock the following morning and reached our destination at ten on Wednesday night. The scenery in many parts is exceedingly grand; twenty-seven miles from the ocean the country is undulating, the most part covered with wood, chiefly pine. On both sides of the river are extensive plains of deep rich alluvial soil with a thick herbage of herbaceous plants. Here the country becomes mountainous, and on the banks of the river the rocks rise perpendicularly to the height of several hundred feet in some parts, over which are some fine waterfalls. The rocks are chiefly secondary, sandstone and limestone bedded on blue granite. The country continues mountainous as far as the lower branch of the Willamette river, the Belle Vue Point of Vancouver about seventy miles from the ocean, when it again becomes low on the banks and rises gradually on the back ground. On the south, towards the head waters of the Williamette, which are supposed to be in a ridge of snowy mountains which run in a south-west direction from the Columbia, the view is fine. A very conspicuous conical mountain is seen in the distance far exceeding the others in height; this I have no doubt is Mount Jefferson of Lewis and Clarke; two others equally conspicuous are observed, one due east and one to the north, the former Mount Hood, the latter Mount St. Helens of Vancouver. Their height must be very great (at least 10,000 to 12,000 ft.), two-thirds are I am informed continually enwrapped in perpetual snow. My residence is on the north bank of the river twelve miles below Point Vancouver (90 from the ocean, the spot where the officer of his squadron discontinued their survey of the river. The place is called Fort Vancouver. In the river opposite my hut lies Menzies Island, so named by Mr Broughton in honour of Archibald Menzies, Esq., then his companion on the famous expedition. On my arrival a tent was kindly offered, having no houses yet built, which I occupied for some weeks; a lodge of deerskin was then made for me which soon became too small by the augmenting of my collection and being ill adapted for drying my plants and seeds. I am now (August 16) in a hut made of bark of *Thuya plicata* which most likely will be my winter lodging. I have been only three nights in a house since my arrival, and three first on shore. In my journeys I have a tent where it can be carried, which rarely can be done; sometimes I sleep in one, sometimes under a canoe turned upside down, but most commonly under the shade of a pine tree without anything. In England people shudder at the idea of sleeping with a window open; here, each individual takes his blanket and with all the complacency of mind that can be imagined throws himself on the sand or under a bush just as if he was going to bed. I confess, at first, although I always stood it well and never felt any bad effects from it, it was looked on by me with a sort of dread. Now I am well accustomed to it, so much so that comfort seems superfluity.

Sunday May 1st Early in the morning, left the fort for the purpose of visiting an extensive plain seven miles below on the same side of the river. Passed several Indian steaming huts or vapour baths; a small hole is dug about 1 foot deep, in which hot stones are placed and water thrown on them so as to produce steam; the bather then goes in naked and remains until well steamed; he immediately plunges into some pool or river, which is chosen so as not to be far distant. They are formed of sticks, mud, and turfs, with a small hole for means of entering. They are most frequently used when the natives come from their hunting parties, after the fatigues of war, and also before they go on any expedition which requires bodily exertion. My curiosity was not so strong as to regale myself with a bath.

June 20th Towards midday left my residence for a journey up the river in company with the canoes going to the different posts in the interior, a few miles above the Great Falls, about two hundred miles from the ocean. I was at a loss to decide whether my time would be better employed there or between here and the ocean. In the latter, from what I have already seen, I should reap a rich harvest, and leave it probably for a less fertile one; although, on the other hand, I might obtain some interesting objects peculiar to the plains and mountains of the interior, as John McLoughlin, Esq., (the chief factor), from whom I have experienced every attention and assistance as to the furthering of my pursuit and comfort which he has in his power to show, assures me there will be no obstacle to my crossing the continent, and that he will use every means to make my journey beneficial to the Society and agreeable to myself. Before the vessel left the river for Nootka and thereabouts, I had some thought of going there. But as he informed me, that my opportunities of collecting, arising from the turbulent disposition of the natives, would be so limited — persons being under the necessity of meeting them armed and in a large party — in unison with his opinion I thought my time would be devoted to the best advantage by remaining on the Columbia, and to make journeys in various directions as opportunities would occur. My ascent was slow, the current at this season being exceedingly powerful, so that I had many excursions on the banks and adjoining hills. The water ran with such rapidity that when the wind blows from a contrary direction it produces a swell like an inland sea; frequently we had to take shelter in the creeks, and although our canoes were considered good, yet we could not see each other except at a short distance, so great was the swell. The Grand Rapids, as they are termed by

Lewis and Clarke, are formed by the river passing through a narrow channel about 270 yards broad in a south-west direction, very rocky, the fall of water about 147 feet above which stand three small islands; one of them is the burial-place of the natives who inhabit the southern banks of the river. The extreme length of the Rapids may be about two miles, but for only a short space (about 600 yards where the river makes a turn S.W.) the water passes with great agitation. At this season they are seen to a disadvantage, the river being 9 feet higher water than in May (from May 24 to July 16 the river rose 12 feet 8 inches); I am informed it is lower this season than generally. The banks are high, steep, and in many places rugged; limestone, sandstone, on blue and grey granite. Many large trees in a petrified state are to be seen lying in a horizontal position between the layers of rock, the ends touching the water in many places. There seem to be two kinds, a soft wood and a hard; one I take to be *Abies balsamea,* the other a species of *Acer,* which must be *A. macrophyllum,* being the only hard wood of large dimensions on the place; some of both measure 5 feet in diameter. This being the fishing season, the natives are numerous on the banks of the river; they come several hundred miles to their favourite fishing grounds. At the Rapids an almost incredible number of salmon are caught. They are taken in the following manner; before the water rises on the approach of summer, small channels are made among the stones and rocks, 2 feet broad and running out into various branches, over which is placed a platform for the person to stand. Several channels are made, some higher, some lower, so as to suit the water as it falls or rises. A scoop net or net fastened round a hoop at the end of a long pole, 12 to 15 feet, is all that is used; the person stands on the extremity of the stage or platform and places his net at the top of the channel, which is always made to fit it exactly, and it is carried down with the current. The poor salmon, coming up his smooth and agreeable road as he conceives it to be, thrusts himself in the net and is immediately thrown on the stage; the handle or pole of the net is tied to the platform by a rope lest the pressure of water or strength of the fish should snatch it out of the hands of the fisher. The hoop is made of *Acer circinnatum* of Pursh, which is very tough and not unlike *A. rubrum.* The pole is balsam pine, which after drying is light. The net is made from the bark of a species of *Apocynum,* which is very durable. The fish are of good quality, much about the same size as those caught in the rivers of Europe, 15 to 25 lb.

generally, some more. I measured two, the one 3 feet 5 inches
from the snout to the extremity of the tail, 10 inches broad at
the thickest part, weighing about 35 lb.; another 3 feet, and
9 inches broad, a little lighter. Both were purchased for 2 inches
of tobacco (¼ oz.) value twopence, or one penny each. How
little the value from that in England, where the same quantity
would cost £3 or £4, and not crisped salmon as it is termed by
those acquainted with refinement of dishes, as I have it, cooked
under the shade of a lordly pine or rocky dell far removed from
the abodes of civilised life. It is wonderful the comfort, at least
the pleasant idea of being comfortable in such a place surrounded
by multitudes of individuals who, perhaps, had never seen a white
person before, and were we to judge by their appearance are
very hostile, viewing us narrowly with surprise. The luxury of
a night's sleep on a bed of pine branches can only be appreciated
by those who have experienced a route over a barren plain,
scorched by the sun, or fatigued by groping their way through
a thick forest, crossing gullies, dead wood, lakes, stones, &c.
Indeed so much worn out was I three times by fatigue and
hunger that twice I crawled, for I could hardly walk, to a small
abandoned hut. I had in my knapsack one biscuit; the third
and last time I was not so bad with hunger, but very weak. I
killed two partridges an hour before I camped, which I placed
in my little kettle to boil for supper. The Canadian and the
two Indians had eaten their dry salmon and were asleep. Before
my birds were cooked Morpheus seized me also; I awoke at
daybreak and beheld my supper burned to ashes and three holes
in the bottom of my kettle. Before leaving my resting place I had
to make a little tea, which is the monarch of all food after
fatiguing journeys. This I did by scouring out the lid of my
tinder-box and boiling the water in it! I have oftentime heard
that 'Necessity has many inventions', which I now know and
partly believe. The natives are inquisitive in the extreme,
treacherous, and will pillage or murder when they can do it
with impunity. Most of the tribes on the coast (the Chenooks,
Cladsaps, Clikitats, and Killimucks) from the association they
have had with Europeans are anxious to imitate them and are
on the whole not unfriendly. Some of them are by no means
deficient of ability. Some will converse in English tolerably well,
make articles after the European models, &c. They are much
prejudiced in favour of their own way of living, although at
the same time will not fail to eat a most inordinate quantity if
offered to them. My canoe-men and guides were much surprised

to see me make an effervescent draught and drink it boiling, as they thought it. They think there are good and bad spirits, and that I belong to the latter class, in consequence of drinking boiling water, lighting my tobacco pipe with my lens and the sun, and they call me Olla-piska, which in the Chenook tongue signifies fire. But above all, to place a pair of spectacles on the nose is beyond all their comprehension: they immediately place the hand tight on the mouth, a gesture of dread or astonishment.

Salmon are also caught on sandy shores, where free from large stones, with a draught net in the same manner as the salmon fishing in Britain. The net is made of *Apocynum* bark, floated by pieces of wood in lieu of cork. This mode is only practised where there are no rapids or projecting rocks, or places to make channels for the scoop net. From the Rapids to the Great Falls, distant about fifty-eight or sixty miles, the banks are steep and in many places rugged. Some of the hills are very high but all destitute of trees or large shrubs. The wood becomes smaller the further the river is ascended. Acer is not founded above the Rapids; *Thuya, Abies balsamea* and one species of *populus* on the edges of creeks, all of which gradually diminish into low scrub-wood. Sixteen miles below the Falls we are no longer fanned by the huge pine stretching its branches in graceful attitude over a mountain rivulet or deep cavern, or regaled by the quivering of the aspen in the breeze. Nothing but extensive plains and barren hills, with the greater part of the herbage scorched and dead by the intense heat. I had to cross a plain nineteen miles without a drop of water, of pure white sand, thermometer in the shade 97°. I suffered much from the heat and reflection of the sun's rays; and scarcely can I tell the state of my feet in the evening from the heat in the dry sand; all the upper part of them were in one blister. Six miles below the Falls the water rushes through several narrow channels, formed by high barren, and extremely rugged rocks about two miles long. It is called by the voyagers the Dalles. On both sides of the river very singular rocks of a great height are to be seen, having all the appearance of being water-worn; not unlikely they have been the boundaries of the river at some former period. The present bed of the river is more than 6000 feet lower. The Falls stretch across the whole breadth of the river in an oblique direction, which may be about 400 yards, about 10 or 12 feet of a perpendicular pitch. At present its effects is somewhat hid, the water being high, but I am told it is fine when the river is low. The ground on both sides is high, destitute of all sorts of wood or shrub except *Berberis nervosa,* and *B. aquifolium, Tigarea tridentate,* and one species of *Ribes* with small red smooth berries.

July 19th Early in the morning I left my residence in a small canoe, with one Canadian and two Indians, for a journey to the shore of the ocean, principally for the purpose of searching for and inquiring after the tuberous-rooted *Cyperus,* the root of which is said to afford the natives food something like potatoes when boiled. After a laborious route of twelve days along the shores north of Cape Disappointment, I was obliged reluctantly to return without being fortunate enough to meet with it. I observed several dead roots, washed on the shore by the surge and agreeing exactly with the description given by Lewis and Clarke, which I conjecture to be it. My guide, who is tolerably conversant with many of the tongues spoken by the inhabitants of the coast, learned that it is very abundant along the shore from Point Adams, the southern entrance of the river, at no great distance. I am for the present prevented from prosecuting my journey in that direction, several of the tribes being at war with each other. I laboured under very great disadvantage by the almost continual rain; many of my specimens I lost, and although I had several oilcloths, I was unable to keep by plants and blanket dry or to preserve a single bird; saw many pelicans of one species, but could not obtain any, (I believe it to be the same as one I killed in the Galapagos,) one albatross, some petrels which did not come under my eye during my voyage out, one large brown gull, and a smaller white with bluish wing on the upper side. This one I have since seen on the sandbanks of the river as far up as the Great Rapids. Now I have a little idea of travelling without the luxuries of life. Only two nights were dry during my stay on the shore; before I could lie down to sleep my blanket drying generally occupied an hour. In the creeks I caught plenty of a small trout and young salmon. With a basin of tea, a small piece of biscuit, and now and then a duck, I managed to live very well. On my return I visited Cockqua, the principal chief of the Chenooks and Chochalii tribes, who is exceedingly fond of all the chiefs that come from King George — words which they learn from Broughton, of Vancouver expedition, and other commanders of English ships. His acquaintance I previously had. He imitates all European manners; immediately after saluting me with 'clachouie', their word for 'friend', or 'How are you?' and a shake of his hand, water was brought immediately for me to wash, and a fire kindled. He then carried me to one of his large canoes, in which lay a sturgeon 10 feet long, 3 at the thickest part in circumference, weighing probably from 400 to 500 lb, to choose what part should be cooked for me. I gave him the preference as to knowledge about the savoury mouthfuls, which he took as a great compliment. In justice to my Indian friend, I cannot but say he afforded me the most comfortable meal I had had for a considerable time before, from the spine and head of the fish. A tent was left here, which could not be carried further, in which I slept. He was at war with the Cladsap tribe, inhabitants of the opposite banks of the river, and that night expected an attack which was not made. He pressed me hard to sleep in his lodge lest anything should befall me: this offer I would have most gladly accepted but as fear should never be shown I slept in my tent fifty yards from the village. In the evening about 300 men danced the war dance and sang several death songs. The description would occupy too much time. In the morning he said I was a great chief, for I was not afraid of the Cladsaps. One of his men, with not a little self-consequence, showed me his skill with the bow and arrow, and then with the gun. He passed arrows through a small hoop of grass 6 inches in diameter, thrown in the air a considerable height by another person; with his rifle he placed a ball within an inch of the mark at the distance of 110 yards. He said no chief from King George could shoot like him, neither could they sing the death song nor dance the war dance. Of shooting on the wing they have no idea. A large species of eagle, *Falco leucocephala*, was perched on a dead stump close to the village; I charged my gun with swan shot, walked up to within forty-five yards of the bird, threw a stone to raise him, and when flying brought him down. This had the desired effect: many of them placed their right hands on their mouths — the token for astonishment or dread. This fellow had still a little confidence in his abilities and offered me a shot at his hat; he threw it up and I carried the whole of the crown away, leaving only the brim. Great value was then laid on my gun and high offers made. My fame was sounded through the camp. Cockqua said 'Cladsap cannot shoot like you'. I find it to be of the utmost value to bringing down a bird flying when going near the lodges, at the same time taking care to make it appear as a little thing and as if you were not observed. In the lodge were some baskets, hats made after their own fashion, cups and pouches, of very fine workmanship; some of them made with leaves of *Typha angustifolia* and leaves of *Helonias tenax;* some with the tissues of *Thuya* roots, and the inner bark, and some with a small linear-leaved *Fucus* with leaves of the stronger *Carices.* I received from him an assemblage of baskets, cups, &c., and his own hat, with a promise that the maker (a little girl twelve years of age, a relation of his own) would make me some hats like the chief's hats from England. I made a short stay, always collecting what came under my notice; on the 5th August I made some small presents of tobacco, knives, nails and gun flints, and then left for my residence up the river to Fort Vancouver, which occupied two and a half days.

August 19th 1825 Towards afternoon left in a small canoe with one Canadian and two Indians, in company with a party of men going on a hunting excursion to southward, on a visit to the Willamette River, one of the southern branches of the Columbia. The distance I was enabled to go was about fifty-six miles. The river is large, nearly as large as the Thames. Thirty-six miles from the Columbia are very fine falls, about 43 feet high, across the whole river, in an oblique direction; when the river is low they are divided into three principal channels, all of which have a perpendicular pier; when the water is high it rushes over in an unbroken sheet. This season in July, which is the time it is at its greatest height, it rose 47 feet. From the Columbia to the Falls there is but little or no current; gorged back by the waters of that river. The banks are covered with *Pseudotsuga menziesii, A. balsamea,* oak and poplar. The soil is by far the richest I have seen. Above the Falls, as far as I went, at many places the current is rapid. I had considerable difficulty in making the portages at the Falls, having to haul the canoe up with ropes; this laborious undertaking occupied three hours, and one hour on my return. This at one time was looked on as the finest place for hunting west of the Rocky Mountains. The beaver now is scarce; none alive came under my notice. I was much gratified in viewing the deserted lodges and dams of that wise economist. Abundance of a species of deer (which probably may be the one spoken of by Mr Sabine in his description of the animals observed by Franklin's party) are to be had. During my stay (ten days) seventeen were killed, both males and females. It grieved me exceedingly I was so placed that only a small one could be preserved: a young male, which

I killed at 115 yards with ball. The flesh is very fine, of a beautiful fine delicate white. Unfortunately I lost my note of their proportions, colour, &c., but as I am very shortly to make a second visit I shall not fail to preserve a pair; horns 3 to 4, branched, short, about 15 inches long; light brown, white on the belly, young ones white spotted until six months old. Near my tent was a small salt-marsh, to which in the morning they daily resorted. Killed 2 females and 3 males of a fine species of pigeon; feet, legs, and part of the beak yellow, a white ring round the neck. I was only able to skin one, a male, and that is a miserable specimen; female same colour except neck and breast, which is of a darker blue. I was very fortunate having good weather all the time, except one day.

Collected the following plants and obtained seeds of several very important plants already collected: among them *Nicotiana quadrivulvis,* correctly supposed by Nuttall to exist on the Columbia; whether its original habitat is here in the Rocky Mountains, or on the Missouri, I am unable to say, but am inclined to think it must be in the mountains. I am informed by the hunters it is more abundant towards them and particularly so among the Snake Indians, who frequently visit the Indians inhabiting the head-waters of the Missouri by whom it might be carried in both directions. I have seen only one plant before, in the hand of an Indian two months since at the Great Falls of the Columbia, and although I offered him 2 ozs. of manufactured tobacco he would on no consideration part with it. The natives cultivate it here, and although I made diligent search for it, it never came under my notice until now. They do not cultivate it near camps or lodges, lest it should be taken for use before maturity. An open place in the wood is chosen where there is dead wood, which they burn, and sow the seed in the ashes. Fortunately I met with one of the little plantations and supplied myself with seeds and specimens without delay. On my way home I met the owner, who, seeing it under my arm, appeared to be much displeased; but by presenting him with two finger-lengths of tobacco from Europe his wrath was appeased and we became good friends. He then gave me the above description of cultivating it. He told me that wood ashes made it grow very large. I was much pleased with the idea of using wood ashes. Thus we see that even the savages on the Columbia know the good effects produced on vegetation by the use of carbon. His knowledge of plants and their uses gained him another finger-length. When we smoked we were all in all.

Returned on the 30th of August. From that time till Thursday, September 1st, employed drying, arranging, putting up seeds, and making up my notes. Early on Thursday went on a journey to the Grand Rapids to collect seeds of several plants seen in flower in June and July. Went up in a canoe accompanied by one Canadian and a Chief (called Chumtalia) of the tribe inhabiting the north banks of the river at the Rapids. I arrived on the evening of the second day and pitched my tent a short distance from the village. I caused my Canadian to drench the ground well with water to prevent me from being annoyed with fleas, although I was not altogether exempt from them, yet it had a good effect. I found my Indian friend during my stay very attentive and I received no harm or insult. He accompanied me on some of my journeys. (They were only a few years since very hostile. The Company's boats were frequently pillaged by them and some of their people killed.) My visit was the first ever made without a guard. On Saturday morning went on a journey to the summit near the Rapids on the north side of the river, with the chief's brother as my guide, leaving the Canadian to take care of the tent and property. This took three days, and was one of the most laborious undertakings I ever experienced, the way was so rough, over dead wood, detached rocks, rivulets, &c. that very little paper could be carried. Indeed I was obliged to leave my blanket (which, on my route is all my bedding) at my first encampment about two-thirds up. My provision was 3 oz. tea, 1 lb. sugar, and four small biscuits. On the summit all the herbage is low shrub but chiefly herb plants. The second day I caught no fish, and at such a great altitude the only birds to be seen were hawks, eagles, vultures, &c. I was fortunate enough to kill one young white-headed eagle, which (then) I found very good eating. I roasted it, having only a small pan for making tea. On the summit of the hill I slept one night. I made a small fire of grass and twigs and dried my clothes which were wet with perspiration and then laid myself down on the grass with my feet to the fire. I found it very cold and had to rise four times and walk to keep myself warm. Fortunately it was dry and a keen north wind prevented dew. On Monday evening at dusk I reached my tent at the village much fatigued and weak and found all things going on smoothly. Made a trip to the opposite side two days later, also to the summit of the hills, which I found of easier ascent, the only steep part near the top. My food during my stay was fresh salmon, without salt, pepper, or any other spice, with a very little biscuit and tea, which is a great luxury after a day's march.

Last night my Indian friend Cockqua arrived here from his tribe on the coast, and brought me three of the hats made on the English fashion, which I ordered when there in July; the fourth, which will have some initials wrought in it, is not finished, but will be sent by the other ship. I think them a good specimen of the ingenuity of the natives and particularly also being made by the little girl, twelve years old, spoken of when at the village. I paid one blanket (value 7s.) for them, the fourth included. We smoked; I gave him a dram and a few needles, beeds, pins, and rings as a present for the little girl. Faithful to his proposition he brought me a large paper of seeds of *Vaccinium ovatum* in a perfect state, which I showed him when there, then in an unripe state. I have circulated notices among my Indian acquaintances to obtain it for me.

Returned on 13th, and on my arrivel I found that Mr Scouler had taken possession of my house and learned that the vessel had returned from the north and would be despatched for England without delay. My time must now be taken up packing, arranging, and writing for a short time. From that time till October 3rd employed dividing my seeds and specimens and finishing transcribing my Journal. Wrote today to Jos. Sabine, Esq., to Dr Hooker and Mr Murray of Glasgow, to A. Menzies, Esq., and to my brother. I am tomorrow morning to leave here to see my boxes safely placed in the vessel. — Fort Vancouver, Columbia River, October 3rd, 1825, D. D.

October 4th - 22nd In consequence of receiving a wound on my left knee by falling on a rusty nail when employed packing the last of my boxes, I am unfortunately prevented from proceeding with my collection to the ship. In the meantime I wrote a note to Captain Hanwell, requesting he would have the goodness to place them in an airy situation, particularly the chest of seed, and, if possible, above the level of the water. I gave him also a note to Joseph Sabine, Esq. He kindly answered my note immediately on receipt and assured me that as far as was in his power he should feel glad in complying with my request, and that he should make a point of calling on Mr Sabine on his arrival in England. On the 7th my leg became violently inflamed and a large abcess formed on the knee-joint which did not suppurate until the 16th. It is needless to observe that I was unable to continue my journeys or increase my collection during the time. This very unfortunate circumstance gave me much uneasiness, being my harvest of seed.

October 22nd - November 15th Learning the ship had been detained by contrary winds, and finding myself much recovered, I left for Vancouver in a small canoe, with four Indians, for the purpose of visiting my old shipmates on my way to Whitbey Harbour on the Cheecheeler River in latitude 48°, near to which were several plants that had not come under my notice or of which I had only obtained imperfect specimens and a supply of seed, among them *Helonias tenax,* a very desirable plant for cultivation. I camped at the junction of the Willamette River at sundown, having made only twenty miles, a strong wind setting in from the sea. On Sunday at daylight I embarked, but before leaving my encampment the canoe had to be fresh gummed. I had not proceeded many miles when it struck on the stump of a tree, which split it from one end to the other, and I had to paddle to shore without loss of time, the water rushing in fast. During the time my Indians were repairing it I occupied the office of cook. I made myself a small basin of tea and boiled some salmon for them. At ten o'clock I proceeded on my route. At eight the same evening I put ashore at the village of Oak Point to procure some food, where an Indian gave me a letter from Mr Scouler, the surgeon of the ship, who informed me in his note they would not yet leave for a few days, and as the vessel was seen that same day in the bay I was desirous of writing to Mr Sabine up to that date. After obtaining a few dried salmon and a wild goose, I went on four miles further down the river, where we took some supper, and continued my journey at ten o'clock, expecting to reach the sea before daylight, being only forty-three miles distant. At four in the morning of Monday a strong westerly breeze set in, which produced a very angry swell on the river and obliged me to cast along the shore. Indeed this was almost necessary under any circumstances, my canoe being so frail. I landed at the mouth of the river at 9 a.m. where I was informed by the Indians the ship had sailed an hour before. I felt no little disappointed, having my letter ready to hand on board. After breakfast my canoe-men lay down to sleep, and I took my gun and knapsack and walked along the bay in quest of some seeds. In the evening, I returned to the lodge of Madsue or Thunder, one of the Chenook chiefs, where I found his brother Tha-a-muxi, or the Bear, a chief from Whitbey Harbour. As he was then going to his home he offered to accompany me, to which I agreed. On Tuesday the 25th Com Comly or Madsue ferried us across the Bay. Our canoe being small, and as I found his so much more commodious, I negotiated with him to lend it to me, which he did in Baker's Bay at the entrance of Knight's River. In the evening I gave the two chiefs a dram of well-watered rum, which pernicious liquor they will make any sacrifice to obtain. I found an exception in my guide Tha-a-muxi; he would not taste any. I inquired the reason, when he informed me with much merriment that some years since he got drunk and became very quarrelsome in his village; so much so that the young men had to bind his hands and feet, which he looked on as a great affront. He has not tasted any since. In lieu of that I found him an expensive companion in the way of smoking; so greedily would he seize the pipe and inhale any particle of smoke in the lungs, that he would regularly five or six times a day fall down in a state of stupefaction. Smoking with them being the test of friendship, it is indispensable. I was of course compelled to join. I found my mode gave him as much sport as his gave me. He observed, "Oh, why do you throw away the food? Look at me, I take it in my belly." On Wednesday I made a portage of four miles over Cape Disappointment, on the north point of the Columbia, to a small rivulet which falls into the ocean twelve miles to the north. I found it very laborious dragging my canoe through the wood, over rocks, stumps, and gullies. On reaching the bay I proceeded along the coast a few miles; two hours before dusk a thick fog with a drizzly rain obliged me to encamp for the night under shelving rock a little above the tide-mark, overshadowed by large pines. In the evening I felt my knee more troublesome and very stiff, arising from the exertion I had to make in transporting the canoe, or probably with the cold and rain. After a comfortless night's rest I resumed my route at daylight, and as I was disappointed in not procuring salmon at the village I passed yesterday, it being abandoned, I had nothing to eat except a small cake of chocolate about two ounces, so with as much speed as possible I proceeded to Cape Foulweather, which I gained in the evening, forty miles being made that day along the coast. As I had here a portage of sixteen miles to make, too great an undertaking to be done by so few, I sent two of the Indians with the canoe to the Columbia. As we had not this day had any food, they preferred leaving that same evening in hopes of obtaining some fish; my guide and the other two remained with me. They had not been away more than two hours when a most violent hurricane set in from the west, producing an agitation in the shoal water frightful in the extreme. I was much alarmed for their safety, but learned on my return they happily put into a sheltered creek at the commencement of the storm and remained till it abated. I was very hungry in the evening, and went out and gathered a few berries of *Arbutus Uva-ursi* being the only thing which could be found at the place. The wind was so high, with heavy rain, that scarcely any fire could be made. Long ere day I was ready to leave Cape Foul-weather, which name it merits; being in a very bad state for walking. All the wildfowl had fled to the more sheltered parts; not a bird of any description could be seen. Being the two days without food, I resolved to endeavour to walk over the portage to the north side of Whitbey Harbour, where I was informed by my guide he expected a fishing party from his village to be. On my arrival there at six o'clock being on my legs from four in the morning, I hardly can give an idea of my afflicted state. The storm continued with equal violence, which prevented the fishing party from leaving their village, which increased my misery. While my guide and the Indians were collecting fuel, I made a small booth of pine branches, grass, and a few old mats; my blanket being drenched in wet the preceding day, and no opportunity of drying it, the night raining heavily, I deemed it prudent not to lie down to sleep. Therefore I spent this night at the fire. On Saturday I found myself so much broken down and my knee so much worse that I did not stir out for the whole of the day. A little before dusk the weather moderated, when I crawled out with my gun; providentially I killed five ducks with one shot, which, as might be expected, were soon cooked; one of the Indians ate a part raw, the other did not take time to pluck the feathers off but literally burned them to save time. I was certainly very hungry, but as soon as I saw the birds fall my appetite fled; it had brought such a change over me that I could hardly persuade myself I had been in want. I made a basin of tea, on which, with a bit of duck, I made a good supper. Very little sufficed me. At midnight my guides arrived; our fire had attracted their notice, and, as the chief was expected, they had come to wait on him. I was asleep, and did not know until Tha-a-muxi roused me in the morning to embark. He would not allow them to wake me or make any noise, having had no sleep last night and very little the two nights before.

Crossing the Bay I killed two large gulls: one white, bluish

on the wings, with black feathers with points; one of an equal
size, of a mottled grey and a species of dove. I had no
opportunity at the time to preserve them. The Cheecheeler River
is a large stream, nearly as large as the Thames, very rapid,
with numerous cascades. I reached my guide's village a little
before dusk, where I had every kindness and all the hospitality
Indian courtesy could suggest, and made a stay of several days
at his house, during which I was fortunate enough to procure
a little seed of *Helonias;* being so late in the season, I was
unable to procure as much as I should have done if earlier.
Abundance of seeds of that splendid *Carex* and *Lupinus* the
roots of which are gathered and roasted in the embers and eaten.
This is the wild liquorice spoken of by Lewis and Clarke. There
is in the root a large quantity of farinaceous substance, and it
is a very meritous wholesome food. I procured several other
seeds not in my possession before. The *Lupinus* is called by
them Somuchtan; seed-vessel one-celled, seed angular; calyx
none; corolla five-petalled, lanceolate; stamens five to nine; style
three-cleft; flowers faint white; leaves alternate, linear, sessile,
revolute; stem suffruticose, covered with chaff scales.

On the 7th November, I proceeded up the river in a canoe
with my guide; made halts at places such as presented anything
different from what I had seen before. On the 11th, I reached
sixty miles from the ocean, where I found my canoe too large
to pass in many places by reason of cascades and shallowness
of the water. I abandoned the idea of proceeding further in that
direction. I therefore made my guide such presents as were
adequate for the service and kindness I had experienced from
him. Before leaving me he requested I would shave him, as he
had pretensions to civilisation and aped with nicety European
manners. I accordingly did so, and invited him to come at the
New Year to see me, when I would give him a dram, a smoke,
and shave him again. He told me before he left, to let all King
George's chiefs know of him, when I spoke to them with
paper. This river is a large stream nearly as large as the Thames,
very rapid in many parts with cascades. The banks are rocky,
steep, and covered with the like woods as are found on the
Columbia. At the village where I put up I bargained with an
Indian to carry my luggage on his horse to the Cow-a-lidsk River,
one of the northern branches of the Columbia. I had some
difficulty with this fellow in accomplishing my end; he was the
most mercenary rascal I have seen. I had to give him twenty
shots of amunition, two feet of tobacco, a few flints, and a little

vermilion. The following day rained so heavily that I could
not proceed. Early on the 13th I set out with my two Indians
on foot, the horse carrying my little baggage with the owner. The
distance may be about forty miles, and a very bad road owing
to the late heavy rains; much water was in the hollows, and
the little creeks and rivulets so much swollen that my clothes
were often off three times swimming across some of them. In
the afternoon the rain fell in torrents, and as the country was
an entire plain and no commodious place for camping, I was
urged to exert myself to endeavour to reach the Cow-a-lidsk,
which I accomplished at sundown, being greatly fatigued. My
track was along the foot of Mount St. Helens of Vancouver,
which lay a little to the north-east. At Schachanaway's or the
chief of the Chenook tribe's house I learned he had just returned
from a trading visit from other parts and had brought with
him a bag of potatoes, flour, a little molasses, and rum, of all
of which I had a portion and a comfortable night's lodging. A
small boat had been lent to him, which I considered fortunate,
as it enabled me to proceed without delay.

On the 14th I had breakfast and was on my route before
five o'clock in the morning. This is a large river, 150 to 200
yards wide in many parts, very deep and rapid, the current
running more than six miles an hour in many parts. At mid-
afternoon camped on a small woody island at its mouth, where
it joins the Columbia, fifty miles from the ocean. Being high
water when I put in, the boat grounded at ebbtide; not having
strength enough to slide her along on the sand, I had to wait
longer in the morning than I would have otherwise done. At
six in the morning of the 15th I proceeded up the Columbia
with a freshening breeze of wind, my blanket and cloak serving
as sails. I arrived again at Fort Vancouver at half-past eleven
at night, being absent twenty-five days, during which I
experienced more fatigue and misery, and gleaned less than in
any trip I have had in the country.

November 16th - December 31st The rainy season being set in,
with my infirm state, totally banished every thought from my
mind of being able to do much more in the way of botany for
season. It is with serious regret that I am compelled to resign my
labours, so much sooner than if that accident had not befallen
me. At midday on the 18th the express, consisting of two boats
and forty men, arrived from Hudson's Bay which they left
on the 21st of July. They were observed at the distance of some

miles, rapidly descending the stream. In this distant land, where there is only an annual post, they were by every person made welcome guests. I hastened to the landing-place, congratulating myself on the news from England. I learned with much regret there were no letters, parcel, or any article for me. I was given to understand they left Hudson's Bay before the arrival of the ship which left London the May before, so that if Mr Sabine wrote to me, the letter will remain on the other side of the continent till next November. I was exceedingly disappointed. A Mr McLeod, the person in charge of the party, told me he met Captain Franklin's party on Cumberland Lake on their way to Bear Lake, their winter residence, early in July; their stay being only a few minutes, Dr Richardson did not write to me.

I learned there was a Mr Drummond attached to them as botanist; he accompanied Mr McLeod as far on his route as the foot of the Rocky Mountains, and is to pass the summer in the country towards Peace and Smoky Rivers. This I take to be Drummond of Forfar, from the description given of him. Mr McLeod, whom I find an agreeable gentlemanly man, and from whom I have had much kindness, informs me that he spent the last five years at Fort Good Hope, on the McKenzie River, and of course possesses more knowledge of that country than any other person; that (if the natives can be believed, with whom he was well acquainted and perfectly conversant in their language) there is a river nearly equal to the McKenzie to the westward of it, running parallel with it, and falls into the sea near Icy Cape. At the mouth of the said river there is a trading establishment on a woody island, where ships come in the summer. The people have large beards and are very wicked; they have hanged several of the natives to the rigging and have ever since been in much disrepute. Much stress may with many be laid on this statement with all safety, as he showed me several articles of Russian manufacture, among which were small copper Russian coins, metal combs, &c.

But the most convincing proof that the difficulty of transportation by land or water is trifling, is large four, five and six gallon malleable-iron pots of very coarse workmanship and very different from anything in the trade of the British Fur Company. He exchanged some of his for theirs. The sea to the west of the McKenzie River is said to be open after July, so that there is little difficulty in going either by water or land to Icy Cape. Mr M. had the Indians assemble for the purpose of extending their territory in that direction, when he had to leave and proceed to Hudson's Bay. In him there is a great example of perseverance, visiting the Polar Sea, the Atlantic and Pacific Oceans, in the short space of eleven months. In the short spells of fair weather, when able, I crawled out, either with my gun collecting birds or other animals, or picking up *Musci* or any Cryptogamic plants in the woods. As yet (15th December) there has been scarcely any frost. When dry, weather generally very pleasant during the day; the nights invariably cold and damp. On the 24th December the rain fell in such torrents, without the least intermission, that my little hut of *Thuya* bark, which stood in rather a low situation, was completely inundated; 14 inches of water was in it. As my lodgings were not of the most comfortable sort, Mr McLoughlin kindly invited me to a part of his house in a half-finished state. Therefore on Christmas Day all my little things were removed to my new dwelling. After the morning service was performed, they took an airing on horseback. I was prevented from joining them in their pleasant excursion by my troublesome knee.

3

1826
The Blue Mountains and
Upper Columbia

The year opens with David Douglas confined to winter quarters in Fort Vancouver. He employed his time observing, writing up notes on the fauna and flora of the region and packing up specimens. On 20th February the hunter Jean Baptist McKay returned from the upper Willamette River bringing with him a massive cone of the hitherto unknown pine, *P. lambertiana* whose huge seeds Douglas had seen in an Indian's shot pouch during the previous summer. Douglas' appetite was whetted.

But his time had nearly expired and he wrote in his Journal:

"From what I have seen in the country, and what I have been enabled to do, there is still much to be done; after a careful consideration as to the propriety of remaining for a season longer than instructed to do, I have resolved not to leave for another year to come. From what I have seen myself of the upper country towards the head-waters of this river and the boundless track contiguous to the Rocky Mountains, I cannot in justice to the Society's interest do otherwise. However, I am uncertain how far I may be justified in so doing. If the motive which induces me to make this arrangement should not be approved of, I beg it may at least be pardoned. In doing so, two considerations presented themselves: first, as I am incurring very little expense; second, being laid up an invalid last autumn during my seed harvest, I lost doubtless many interesting things which I would have otherwise had. Lest the former should be made any objection to, most cheerfully will I labour for this year without any remuneration, if I get only wherewith to purchase a little clothing."

In fact 1826 was to be a fantastic year, for Douglas travelled nearly 4,000 miles.

By early July he had got up the Columbia as far as Kettle Falls — near present day Colville, travelled down to Spokane House to get his gun repaired and made two journeys into the Blue Mountains. On his way up the Kettle Falls he discovered the pretty Mission-Bell — *Fritillaria pudica* and at Spokane an important new conifer, *Pinus ponderosa,* was added to his list.

Sunday January 1st Commencing a year in such a far removed corner of the earth, where I am nearly destitute of civilised society, there is some scope for reflection. In 1824, I was on the Atlantic on my way to England; 1825, between the island of Juan Fernandez and the Galapagos in the Pacific; I am now here, and God only knows where I may be the next. In all probability, if a change does not take place, I will shortly be consigned to the tomb. I can die satisfied with myself. I never have given cause for remonstrance or pain to an individual on earth. I am in my twenty-seventh year.

Monday March 20th On the afternoon of Monday, the 20th, at four o'clock, I left Fort Vancouver in company with John McLeod, Esq., a gentleman going across to Hudson's Bay, and Mr Francis Ermatinger, for the interior, with two boats and fourteen men. The day was very rainy, and we camped on a low piece of ground among poplars and willows, on the north side of the river, a few miles from the establishment, at dusk. The following morning at daylight we proceeded up the river. As there was a strong easterly wind against us, we only gained thirty-five miles; camped seven miles below the Grand Rapids; continued rain throughout. The following day made a portage over the Rapids and camped on a small stony island ten miles above them. Showery. At this season the Rapids are seen to advantage, the river being low. The scenery at this season is likewise grand beyond description; the high mountains in the neighbourhood, which are for the most part covered with pines of several species, some of which grow to an enormous size, are all loaded with snow; the rainbow from the vapour of the agitated water, which rushes with furious rapidity over shattered rocks and through deep caverns producing an agreeable although at the same time a somewhat melancholy echo through the thick wooded valley; the reflections from the snow on the mountains, together with the vivid green of the gigantic pines, form a contrast of rural grandeur that can scarcely be surpassed.

Thursday 23rd Having a strong westerly wind, we proceeded on our journey at daylight under sail and reached the lower part of the Dalles at dusk 6 miles below the Great Falls; camped in a small cove, under a shelving rock. Fortunately the night was fine and pleasant, clear moonlight, which was the more agreeable as our tent could not be well pitched. As the natives had collected in greater numbers than we expected, and showed some disposition to be troublesome on not getting such a large present of tobacco as they wanted, we were under the necessity of watching the whole night. Having a few of my small wax-tapers still remaining, which I lay great value on, I wrote a short note to Mr Murray at Glasgow.

Friday 24th After a tedious night, daybreak was to me particularly gratifying, as might be well guessed, being surrounded by at least 450 savages who, judging from appearances, were everything but amicable. As no one in the brigade could converse with them better than myself, little could be done by persuasion. However, finding two of the principal men who understood the Chenook tongue, with which I am partially acquainted, the little I had I found on this occasion very useful. We took a little breakfast on the rocks at the Dalles, four miles below the Great Falls, at seven o'clock. The day was very pleasant, with a clear sky. At five in the evening we made the portage over the Falls, where we found the Indians very troublesome. I learned from Mr McLeod they had collected for the purpose of pillaging the boats, which we soon found to be the case. After they had the usual present of tobacco, they became desirous of our camping there for the night, no doubt expecting to effect their purpose. The first thing that was observed was their cunningly throwing water on the gun locks, and on the boats being ordered to be put in the water they refused to allow them. As Mr McLeod was putting his hand on one of their shoulders to push him back, a fellow immediately pulled from his quiver a bow and a handful of arrows, and presented it at Mr McLeod. As I was standing on the outside of the crowd I perceived it, and, as no time was to be lost, I instantly slipped the cover off my gun, which at the time was charged with backshot, and presented at him, and invited him to fire his arrow, and then I should certainly shoot him. Just at this time a chief of the Kyeuuse tribe and three of his young men, who are the terror of all other tribes west of the mountains and great friends of the white people, as they call them, stepped in and settled the matter in a few words without any further trouble. This very friendly Indian, who is the finest figure of a man that I have seen, standing nearly 6 feet 6 inches high, accompanied us a few miles up the river, where we camped for the night, after being remunerated by Mr McLeod for his friendship — I being King George's Chief or the Grass Man, as I am called. I bored a hole in the only shilling I had, one which has been in my pocket since I left London, and, the septum of his nose being perforated, I suspended it to it with a brass wire. This was to him the great seal of friendship.

After smoking, he returned to the Indian village and promised that he would not allow us to be molested. Of course no sleep was had this night, and to keep myself awake I wrote a letter to Dr Hooker. Heavy rain during the night. The following day, the 25th, at daylight we resumed our route; sleet and rain, with a keen north wind. Being almost benumbed with cold, I preferred walking along the banks of the river, and, although my path in many places was very rugged, I camped forty miles above the Falls much fatigued. During the night and the following morning I found my knee troublesome and very stiff.

26th - 28th At three o'clock on the 28th arrived at Wallawallah establishment, where I was very friendly recieved by S. Black, Esq., the person in charge. The whole country from the Great Falls to this place is nearly destitute of timber. Dry gravelly and rocky soils, with extensive plains. The largest shrub to be seen on the plains is *Tigarea tridentata,* which we invariably used as fuel in boiling our little kettle, also several very curious species of shrubby *Artemisia,* and other shrubs which to me were perfectly unknown; and the whole herbage very different indeed from the vegetation on the coast. To the south-east, at the distance of ninety miles, is seen a ridge of high snowy mountains, which run from north-east to south-west and terminate near the ocean, about 300 miles south of the Columbia; this place will afford very likely most of the plants found in the chain of the Rocky Mountains. Mr Black has very kindly made arrangements for my journey early in June, which will at least occupy 15 to 20 days. The course of the river from this place to the ocean is south-west, many places are very rapid, not more than 50 to 70 yards, which renders it very dangerous.

Early on Thursday morning the 30th proceeded on our route. As the whole country was an extensive plain, I walked on the north side the river till ten o'clock, when we stopped for breakfast, opposite to Lewis and Clarke's River (the Snake River), a stream of considerable magnitude, 100 to 150 yards wide at many parts and likewise rapid. Salmon, I learn, are caught in great abundance as far up as the Falls, and on some of its branches in the immediate vicinity of the Rocky Mountains, passing through a tract of country not less than 1500 miles.

April 1st Here it becomes mountainous, of white clay, with scarcely a vestige of herbage of verdure to be seen, except in the valleys. The river here is much broader than lower down and makes a great band running due east-south, parallel with the coast, and south-east. Camped on the Priest Rapids at seven o'clock in the evening. The river here is narrow, divided into two channels, with a narrow dall through the small rocky island in an oblique direction. The rocks are very rugged, of limestone, and this is considered one of the most dangerous parts of the whole river. During the time of making the portage of nine miles I wrote to my old companion, Mr Scouler of Glasgow.

Monday and Tuesday 10th and 11th Arrived at the junction of the Spokane River with the Columbia at sunset, where we found John Warren Dease, Esq., commandant in the interior, and a party of fourteen men, on their way to the Kettle Falls, ninety miles further up the Columbia. I was by this gentleman received with extreme kindness and had every attention and kindness that could add to my comfort. This is a brother of the gentleman now accompanying Captain Franklin on his two journeys to the Polar Sea. Mr Dease, to whom I was made known through the general notice sent by that agreeable gentleman Mr McLoughlin, at Fort Vancouver, gave me greater hopes than ever of making a rich harvest. He will do all in his power to assist me. This part of the Columbia is by far the most beautiful that I have seen; very varied extensive plains with groups of pine-trees, like English lawn, with rising bluffs or little eminences covered with small brushwood, and rugged rocks covered with ferns, mosses, and lichens.

Thursday 13th April Busy copying the remainder of my notes, as Mr McLeod is to leave early in the morning for his long trip to Hudson's Bay. I am particularly obliged to this gentleman for his friendly attention. He has in the most careful manner taken my small tin box of seeds in his own private box and will hand it over to Mr McTavish. He has also taken my package of notes. I met here Mr John Work, with whom I was acquainted last year, and who sent me a few seeds from the interior last November, and furnished me with some valuable information about the plants and mountain sheep in this neighbourhood.

Wednesday 19th April On Wednesday at eleven o'clock, in company with Mr Dease and his party of fourteen men and two boats, I left this place for the new intended establishment called Fort Colville, near the Kettle Falls, ninety miles further up the river. I am much indebted to this gentleman for the care he took in placing my paper and other articles in a safe place in the boat, and for the kindness he showed myself by inviting me to a seat in his own boat. The whole distance is very mountainous and rugged, the nearer to the Rocky Mountains more so, and more thickly wooded, of three species of conifer. One *P. resinosa* (?); *P.* not unlike *menziesii* found near the coast, but by no means attains such a size; larch is found in abundance in the mountain valleys, much larger than any I have seen on the other side of the continent or even read of. I measured some 30 feet in circumfrence, and several that were blown down by the late storms 144 feet long; wood clean and perfectly straight. On the plains and valleys there is a thick sward of grass, and interspersed among the detached rocks are several species of shrubs which at this season I cannot ascertain; the greater part of the hills covered with snow. Warm during the day, keen frosts at night; maximum 65°, minimum 28°. Camped on the channel of the river as no place could be found more suitable, having made twenty-seven miles; river very rapid. At 4 a.m. on Thursday raised camp and proceeded on our route prosperously throughout, having gained forty miles; the whole distance I walked on foot except being crossed three times, as I could not pass by steep rocks. On Friday at daylight continued our journey, and as we had gained a very rapid place where a portage had to be made, we took breakfast a little earlier than usual, being nine o'clock. This rapid, which nearly equals the Grand Rapids, 150 miles from the ocean, having no name, I called it Thomson's Rapids after the first person who ever descended the whole chain of the river from its source to the ocean. About ten o'clock it began to rain heavily and continued so until four in the afternoon. Arrived at the Falls at six in the evening, thoroughly drenched to the skin, and gladly walked over the portage three-quarters of a mile to a small circular plain surrounded by high hills on all sides, where the new establishment is to be. After our tents were pitched we had a comfortable supper of salmon trout, and dried buffalo meat served up to us by the man who started the day before us with a band of horses. Although my plants were covered with a double oilcloth, I found it inadequate to keep them dry, and, lest any should be injured, such as were wet I put in dry paper, and placed under some pieces of bark near the fire for the night.

Tuesday 9th May 1826 Left the Kettle Falls on the Columbia River at 10 a.m. with two horses, one carrying my provisions, which consisted of buffalo dried meat, a little tea and sugar, my blanket and paper; the other for carrying me over the bad places of the way. I had for my guides two young men, sons of of Mr Jaques Raphael Finlay, a Canadian Sauteur, who is at present residing in the abandoned establishment of Spokane, in which direction I was going. Mr Finlay being a man of extensive information as to the appearance of the country, animals, and so on, Mr Dease kindly gave me a note to him requesting that he would show me anything that he deemed curious in the way of plants, &c. Took my departure in a northerly direction over the mountains, towards the Spokane River, distant about 100 or 110 miles. As my path, running along the skirt of the mountains was at this season very bad and scarcely passable, the numerous mountain rivulets being so much swollen by the melting of the snow, and the meadows being overflowed or so soft that the horses could not pass, obliged me to make a more circuitous route than if it had been later in the season. Camped under a large pine on a rising bluff in the centre of a large plain at four o'clock, having made about twenty-seven miles. After hobbling the horses, took a walk around my camp.

Wednesday 10th May Rose at daylight and had my horses saddled, and being desirous of making the most of my time I took no breakfast further than a little dried meat and a drink of water, and proceeded on my journey at five o'clock. At twelve noon reached a small rapid river called Barrière River by my guides, which took up an hour in crossing. As there were no Indians near the place, we had to choose either making a raft or to swim. As the latter was the easier method, and all of us good water-men, we unsaddled the horses and drove them in. They all went over well except the last, which entangled itself by the hind legs among some brushwood and struggled much for a considerable time; fortunately the wood gave way and he reached the shore much better than I had any reason to expect. I made two trips on my back, one with my paper and pen, the other with my blanket and clothes — holding my property above water in my hands. My guides made three trips each with the saddles and provisions. Breadth of river 30 yards; heat of the water 40°. During this time there was a very heavy shower of hail, and being nearly half an hour in the water I was so much benumbed with cold that I was under the necessity of kindling a fire. After handing my guides a pipe of tobacco and making ourselves comfortably warm, I continued my route through a delightful undulating country till three o'clock, when I began to ascend a second ridge of mountains which I crossed and camped at dusk at their base in a thick woody valley near a small stream of water on the dry rocky ground. The small beautiful species of Phlox which I found some time since on the Columbia gave the whole open places a fine effect. Flower changeable, white, blue, and fine pink colours.

Thursday 11th Heavy rain during the night, which roused me long ere day. In the twilight of the morn I raised camp, the weather assuming a more inviting appearance. At seven in the morning gained the summit of the last range of hills between the two rivers, and had one of the most sublime views I ever beheld. As I approached the banks of the Spokane River the soil became more barren, except small belts of low ground in the valleys — near the mountain rills. Reached the old establishment at Spokane at eleven o'clock, where I was very kindly received by Mr Finlay. He regretted exceedingly that he had not a single morsel of food to offer me. He and his family were living for the last six weeks on the roots of *Phalangium quamash* (called by the natives all over the country Camass) and a species of black lichen which grows on the pines. The manner of preparing it is as follows: it is gathered from the trees and all the small dead twigs taken out of it, and then immersed in water until it becomes perfectly flexible, and afterwards placed on a heap of heated stones with a layer of grass or leaves between it and the stones to prevent its being burned; then covered over with the same material and a thin covering of earth and allowed to remain until cooked, which generally takes a night. Then before it cools it is compressed into thin cakes and is fit for use. This process is similar to the preparing of *Phalangium*. A cake of this sort and a small basin of water was all he had to offer me. By the kindness of Mr Dease, I had ample provision for fourteen days, with a good stock of game in the saddle-bags which I killed on my way, and this enabled me to share the half of my stock with him; such fare as I had, although very palatable, cannot be considered fine living, but was to him the best meal he had enjoyed for some time. As the principal object of my journey was to get my firelock arranged by him, being the only person within the space of eight hundred miles who could do it, and being an item of the utmost consequence to have done soon, I lost no time in informing him of my request. Unfortunately he did not speak the English language, and my very partial knowledge of French prevented me from obtaining information which I should have acquired. In the afternoon I made a walk up the river and returned at dusk, when I found he had obligingly put my gun in good order, for which I presented him with a pound of tobacco, being the only thing I had to give.

Saturday 13th May As I thought of bending my steps again towards the Columbia, Mr Finlay offered that one of his sons should escort me, which I accepted. Before parting with him I made inquiry about a sort of sheep found in this neighbourhood, about the same size as that described by Lewis and Clarke, but instead of wool it has short thick coarse hair of a brownish-grey, from which it gets the name of Mouton Gris of the voyageurs. The horns of the male, of a dirty-white colour, from a volute sometimes weighing 18 to 24 lb. The horns of the female are bent backwards and curved outwards at the point, about 10 inches or a foot long. The flesh is fine. I offered a small compensation to the sons to procure me skins of male and female, at the same time showing them what way they should be prepared. He assured me that in all probability he would be able to find them about August, as he was going on a hunting trip to the higher grounds contiguous to the Rocky Mountains. Close to the old establishment an Indian burial-ground is to be seen, certainly one of the most curious spectacles I have seen in the country. All the property of the dead, consisting of war implements, garments, gambling articles, in fact everything. Even the favourite horse of the departed is shot with his bow and arrow, and his skin with the hoofs and skull hung over the remains of deceased owner. On trees around the ground small bundles are to be seen tied up in the same manner as they

E

tie provisions when travelling. I could not learn if this was as food or as a sacrifice to some of their deities. The body is placed in the grave in a sitting position, with the knees touching the chin and the arms folded across the chest. It is very difficult to get any information on this point, for nothing seems to hurt their feelings more than even mentioning the name of a departed friend. Left Spokane at 8 a.m. with one guide; went on the same track that I came. As I saw nothing different on my way from what I had previously observed, my stoppages were fewer; gained Barrière River a little before dusk, which I crossed in the same way as I did a few days before; heavy rain all the afternoon, camped a few yards from the banks of the river, in the shade of some pines.

Sunday 14th Very rainy during the whole night; although tolerably well sheltered and had a large fire to sit at, yet I felt cold, my blanket and clothing being wet. As I could not sleep I rose at two o'clock and with some difficulty dried my blanket and a spare shirt, in which I placed my paper containing the few plants collected. Afterwards boiled my small kettle and made some tea. Felt a severe pain between my shoulders, which I thought might arise from the cold in swimming and lying in wet clothes. Therefore, as I had no medicine to take, I set out a little before 4 a.m. on foot, driving the horses before me, thinking that perspiring would remove it, which it partly did. On arriving at my first night's encampment at midday, I stopped a short time to look for the currant in perfection which I saw on my way out just coming to blossom, and fortunately found it in a fine state.

Reached the Kettle Falls on the Columbia in the evening, and although I have not obtained a great number of plants, yet with the repairing of my gun and the few plants collected I must say I felt satisfied.

Between May 15th and June 4th Douglas based himself on Kettle Falls and made many day expeditions into the surrounding country, including a canoe trip up the hitherto unexplored Dease River. He found the region very rich botanically and collected three large bundles of dried plants besides the skins of mountain or rock grouse, curlews and pheasant.

June 5th Rose at half-past two o'clock and had all my articles given over in charge to Mr Dease. My tent struck and breakfast before five, when I took my leave, in company with Mr William Kittson, of the wild romantic scenery of Kettle Falls. We went on horseback two miles from the new establishment, where the boats had been laid up, and embarked at seven promptly. The river by the melting of the snow is much swollen, fully twelve or fourteen feet where it is six hundred yards wide; on getting into the current the boats passed along like an arrow from the bow. Half an hour took us to Thomson's Rapids, where, as I observed on my ascent, the water is dashed over shattered rocks, producing an awful agitation of the water from side to side. On being visited by Mr Pierre L'Etang, the guide, he observed the water was in fine order for jumping the Rapids, as he termed it. Good as it appeared to him, I confessed my timidity prevented me from remaining in the boat. Although I am no coward in the water and have stood unmoved, indeed with pleasure, at the agitation of the ocean raging in the greatest pitch, yet to descend such a place I can never do unless necessity calls for it. Therefore Mr Kittson and I walked along the rocks. No language can convey an adequate idea of the dexterity shown by the Canadian boatmen: they pass through rapids, whirlpools, and narrow channels where, by the strength of such an immense body of water forcing its way, it is risen in the middle to a perfect convexity. In such places, where you think the next moment you are to be dashed to pieces against the steep rocks, they approach and pass with an indescribable coolness, leaving it behind cheering themselves with an exulting boat-song. Reached the junction of the Spokane River twenty minutes after three o'clock of the same day, having in the short space of eight hours made a distance of ninety miles, which may give some idea of the current. As Mr K. made a stay of an hour, securing some boats from the sun, I took a walk around our old April encampment, when I added several items to the collection. Camped at dusk opposite the Cinqpoil River, forty miles further down on the south side of the river. I was much obliged by Mr K. kindly putting to shore for anything that attracted my attention.

Tuesday 6th Embarked a few minutes before three and continue dour route; passed the Little Dalles at eleven, where I walked over the rocks and, as usual, gleaned something for my collection. At one, arrived at Okanagan establishment, where I found my old friend Mr John Work, William Conolly, Esq., a Mr P. C. Pambrun and a James Douglas, with a party of men from Western Caledonia, and a Mr Francis Ermatinger from Thomson's River, a brother of the young man who accompanied me in the spring, all on their way to Fort Vancouver. I shall ever feel no small degree of pleasure on thinking of the kindness I had from these people, which is naturally doubly esteemed in this distant uninhabited country. I must mention in particular the genuine and unaffected friendliness of Mr Conolly, who instantly begged that I would consider myself as an old acquaintance. Mr Work, whom I have so often spoken of, kindly preserved for me a large female grouse and a male black rock-grouse, both very well done, with a few eggs of the former; but as my time was so much taken up collecting plants and changing the paper of those gathered on my way down, I had no time to make a box to place them in, so I left them until the autumn with Mr Ermatinger. Made a turn round the rocks west of Okanagan River, and again made some more additions.

Wednesday 7th At eight this morning with a brigade of five boats I left for Wallawallah, at the junction of the Snake River with the Columbia, which I intend to make my head-quarters for six or eight weeks. Passed the Stony Islands, a place in the river about half a mile in length, exceedingly rugged and dangerous. At four and shortly afterwards, camped on the south side of the river, earlier than usual, two of the boats having been broken. This circumstance gave me a few hours among the rocks, which I spent to great advantage. Killed a large rattle-snake, 3 feet long, on the edge of the river among some stones. Thermometer at noon 92° in the shade. At night the heavens appeared an entire sheet of lightning, until midnight, without thunder or rain.

Thursday 8th As usual started at daylight and took breakfast below the Priest Rapids, on some very fine fresh salmon and buffalo tongue. As Mr Conolly was very desirous of reaching the next establishment this night, no time could be lost. He informed me that his stay at that place would be very short, therefore in the afternoon I wrote to Joseph Sabine, Esq.; but as a strong south wind with heavy rain began about five o'clock, obliged me to leave my letter in a half-finished state. Arrived at Wallawallah, where I was kindly received by Mr Black. Having had very little sleep since I left Kettle Falls, I thought of indulging in six or seven hours at least, so I laid myself down early on the floor of the Indian Hall, but was very shortly

afterwards roused from my slumber by an indescribable herd
of fleas, and had to sleep out among the bushes; the annoyance
of two species of ants, one very large, black, ¼ of an inch long,
and a small red one, rendered it worse, so this night I did not
sleep and gladly hailed the returning day. As soon as I could
see to make a pen I finished the letter to Mr Sabine.

Afterwards wrote to Mr McLoughlin and all my kind
friends at Fort Vancouver, respecting my articles going by ship.
Mr Work obligingly undertook to see my box and articles landed
safely, and Mr Conolly handsomely presented me with 12 feet
of tobacco (better than 2 lbs.) to assist me in my travels, during
their absence at the sea. Being as it were the currency of the
country and particularly scarce, I esteemed it invaluable, as
it will enable me to have guides and other services performed
more willingly.

Thermometer in the shade 90°; minimum 63°. Having had
little sleep for the last five nights, I felt somewhat fatigued. I
went to bed earlier than usual, and shortly after dusk an Indian
arrived from Fort Vancouver with news of the arrival of a ship
in the river, and brought me letters and a small parcel of news-
papers. I grasped the parcel eagerly and tore it open, turning
over my letters; at last I found one in Mr Goode's handwriting;
all was right. I thought one was from Mr Sabine, but on opening
I found it from Mr G. Immediately opened my note from Mr
McLoughlin, who informed me he had kept what he conceived
to be letters from the Society until the people return, not deem-
ing it prudent to risk them by the Indian. I had one from Mr
William Booth, which with Mr G's is all that have as yet come
to hand. Never in my life did I feel in such a state; an uneasy,
melancholy; but pleasing sensation stole on my mind, with an
inordinate longing for the remaining part, and although I did
not hear directly from my friends, I now for once in my life
enjoy and relish the luxury of hearing from England. I had
letters from all my kind friends on the coast, full of expressive
hopes that my labours might be amply rewarded. It is a circumstance worthy of notice that I should write to England in the
morning and receive letters on the same day, for in this uninhibi-
ted distant land the post calls but seldom. The express for the
coast with my letters had only left me six hours when the Indian
arrived.

Saturday June 10th Morning cloudy; warm last night. I pored
over my two letters till after midnight, when I lay down on my

mat. It is needless to say, although I had not slept twelve hours for the five preceeding nights, this one I passed without as much as closing my eyes. About noon I felt very unwell and did not deem it prudent to make much exertion during the intensive heat of the day. Therefore I employed myself mending some of my shoes and walked along the river in the afternoon.

Friday June 16th Wind more moderate, warm and pleasant; making preparations for a journey to the south of the Blue Mountains and Grande Ronde, distant about 140 miles. Sent to the Indian camp to inform my guide to be ready tomorrow at sunrise. Last night I was much annoyed by a herd of rats, which devour every particle of seed I had collected, cut a bundle of dry plants almost right through, carried off my razor and soap-brush. One, as he was in the act of depriving me of my inkstand, which I had just been using before I lay down and was lying close to my pillow, I lifted my gun (which is my night companion as well as day, and lies generally alongside of me, the muzzle to my feet) and gave him the contents. I found it a very strange species, body 10 inches long, tail 7; hairy belly nearly white, back light brown, point of the hairs darker; ears very large, ¾ of an inch long; whiskers jet black, 3 inches; long nose, pointed. In the hurry to recover my inkstand, and the great desire that I had of securing him, I did not take time to change the shot for a smaller sort. Rose early, soon after daylight, and watched; I had not sat up above half an hour when a second came. I handed it a lighter shot and did not destroy the skin. In every respect like the male, only about an inch shorter, the head smaller, body as thick; was not with young. I am informed they are found in great abundance in the Rocky Mountains particularly to the north of the Peace and McKenzie Rivers, where in the winter they do much injury to everything that comes in the way.

Saturday 17th My guide did not arrive from the camp until 8 a.m. and I was uncertain if he would come that day, the horses were not brought in from the meadow, nor my provisions put up. Considerable time was taken up explaining to him the nature of my journey, which was done in the following way; I told Mr Black in English my intended route, who translated it to his Canadian interpreter, and this person communicated it to the Indian in the Kyuuse language, to which tribe he belongs. As a proof of the fickle disposition and keenness of bargain making in these people, he made without delay strict inquiry what he should get for his trouble. This being soon settled, then came the smaller list of present wants, beginning, as his family had been starving for the last two months, and he going just at the commencement of the salmon season, by asking Mr Black to allow them something to eat should they call, which was promised. Afterwards a pair of shoes, and, as his leggings were much worn, leather to make new ones was necessary; a scalping knife, a small piece of tobacco, and a strip of red coarse cloth to make an ornamental cap. This occupied two hours and was sealed by volumes of smoke from a large stone pipe. Mr B. kindly offered to send a boy twelve years of age, the son of the interpreter, who speaks the language fluently, with me, which I gladly accepted. As he spoke a little French, I would be the better able to make known my wants to my guide. I had provided for me three excellent horses for carrying my paper, blanket, and provisions, which was equally divided, and as I choose to walk except on bad places of the road or crossing the creeks, I placed a little more on my horse. Mr B. had put up for me a supply of pemmican, a little biscuit, sugar, and tea, and I was amply supplied for ten days with dry salmon for the guides; with what we would kill I might consider myself comfortable. Towards midday I took my leave in a south-easterly direction, along the banks of Wallawallah River, which I crossed a short distance from its junction with the Columbia. Proceeded slowly along on the south side of the river, always making halts collecting anything different from what I had seen. As water is very scarce in crossing the plains, and the day being far spent I camped on the edge of a small spring among some birch.

Sunday 18th I was dreadfully annoyed with mosquitoes during the night and was roused at two in the morning by a heavy shower of rain which lasted till half-past four. I had the horses brought in and started at five, and continued my journey. This morning I found great relief, the atmosphere being cool and the sand prevented from blowing. This part of the country to the north is an entire level plain of gravel and sand, destitute of timber, not even a shrub exceeding 4 feet in height, except a few low straggling birch and willows on the sides of rivulets or springs.

Tuesday 20th June 1826 As usual started at daylight with a view of reaching the height of land against dusk. The further I went the more difficult I found my undertaking. At midday I made a short stop, where I passed the first snow and collected several plants. Immediately after eating a little dried salmon and a mouthful of water from a chilly crystal spring, I continued my route until 4 p.m. where the horses were stopped by deep wreaths of eternal snow, about 1500 feet below the extreme height of the range. As my object was, if possible, to reach the low alluvial grounds on the opposite side, where I had great expectations, my disappointment may be imagined. However, in the meantime I selected my camp under a projecting rock, saw the horses hobbled, and as it appeared to me my guide seemed somewhat alarmed, I thought it prudent to give him a little time to cool or change his opinion. Therefore I set out on foot with my gun and a small quantity of paper under my arm to gain the summit, leaving them to take care of the horses and camp. In the lower parts I found it exceedingly fatiguing walking on the soft snow, having no snowshoes, but on reaching within a few hundred feet of the top, where there was a hard crust of frost, I without the least difficulty placed my foot on the highest peak of those untrodden regions where never European was before me. The height must be great — 7000 to 7500 from the platform of the mountain, and on the least calculation 9000 above the level of the sea. (Thermometer at 5 p.m. 26° Fahr.) Two days before the maximum heat at the foot of the mountains was 92°, and this day I have every reason to think was equally warm. The view of the surrounding country is extensive and grand. I had not been there above three-quarters of an hour when the upper part of the mountain was suddenly enveloped in dense black cloud; then commenced a most dreadful storm of thunder, lightning, hail, and wind. I never beheld anything that could equal the lightning. Sometimes it would appear in massy sheets, as if the heavens were in a blaze; at others, in vivid zigzag flashes at short intervals with the thunder resounding through the valleys below, and before the echo of the former peal died away the succeeding was begun, so that it was impressed on my mind as if only one. The wind was whistling through the low stunted dead pines accompanied by the merciless cutting hail. As my situation was not a desirable one for spending the night, and it was creeping on me, I hastily bent my steps to my camp below, which I providentially reached at eight o'clock, just in the twilight, the storm raging still without the least appearance of abating. The horses were so alarmed that I found it necessary to tie them to some trees close to the camp. As no fire could be kept in, my supper was of the same quality as my breakfast; and as all my clothes were wet, and having nothing to change, I stripped and rolled myself in my blanket and went soundly to sleep shortly afterwards. Precisely at twelve I was so benumbed with cold that on endeavouring to get up I found my knees refused to do their office. I scoured them well with a rough towel, and as the storm was over made a cheering fire. I could not resist the temptation of making a little tea, which I found restored me greatly (thermometer 26°). If I have any zeal, for once and the first time it began to cool. Hung my clothes up to dry and lay down and slept until three o'clock.

Wednesday 21st   I found the spirits of my guide and interpreter greatly damped, and as I had not abandoned the idea of gaining the other side by a more circuitous and less elevated route, a little more to the south than my present situation, I found their fear increasing; and, as I had every reason to suppose, the young rascally boy told the Indian the reverse of what I wanted him to do. At last the Indian told me that I might be able to go but as I had a river to cross (the Utalla), which at this season was very large, I would require to swim or make a raft; and as his nation was at war with the Snake tribe and we going on the confines of their lands, in all likelihood they would steal our horses and perhaps kill us. As it would be very improper to force him to go with me, and impossible for me to go alone, I was reluctantly obliged for the present to give up the idea of crossing. Therefore I lost no time in gleaning on this side. Camped half way between my Monday's and Tuesday's encampments.

Monday 26th   As I had still time to spare and being somewhat disappointed on my trip last week, I thought of making a second journey to the same mountains in a different direction. On my guide being told last night to be ready at daybreak, he instantly began to complain that the fatigue of the former journey had weakened him so much that he did not think he would be able to proceed. Today he did not make his appearance till 6 o'clock a.m. when he refused to go. He was certainly a little broken down, but I saw nothing that would seriously injure him; and I was on the eve of giving him a little corporal chastisement to teach him that I was not to pay him and not to have his services. He lost no time in making his escape. I learned that the Young Wasp, the interpreter's son, who accompanied me, had told him I was a great medicine man, which is always understood as a necromancer, or being possessed of or conversant with evil spirits, and had the power of doing great wonders; and should he go with me, if he did not do as I wanted, though very likely I did not kill him, he might depend I would turn him into a grisly bear to run and live in the mountains, and he should never see his wife again, which of course acted powerfully on him. The boy was paid by his father according to the merit of his services. I was detained till ten o'clock, when Mr Black procured me another guide, whom I took the more willingly as he was no smoker and at the time such a knave, that no one would dare to steal from him; and it is worthy of notice that there has hardly ever been an instance of dishonesty known when trust was placed in them by depositing property in their hands. Proceeded along the north banks of the Walla-wallah River, and then took to the north branch, intending to touch on the range of mountains a little more to the north from where I was last week. As my present journey is through the same sort of country and climate as the preceding, I found until I reached the mountains, which occupied better than two days, but few plants that I had not seen before. I was more fortunate than before, as the weather continued dry but oppressively warm. Maximum heat 84° to 98°; minimum 64°, and on the mountains 31°.

Sunday July 9th 1826 At the junction of the Snake River. Wrote as follows to Jos. Sabine, Esq.: —
Dear Sir, — This day month I wrote you from this place, and at that time I stated how my time would be taken up during the summer. I have a few days ago arrived from a fatiguing journey on the Blue Mountains, spoken of in my last letter, and have been very successful. I have found on those alpine snowy regions a most beautiful species of *Paeonia, Lupinaster macrocephalus,* a splendid species of *Trifolium* equally fine, *Lupinus argenteus* of Pursh, and another species by far the finest of the tribe, not even excepting *L. nootkatensis;* has a spike a foot to 20 inches of full blossom of a deep golden-yellow; I have placed Mr Turner's name behind it; the plant 4 to 6 feet high. One species of Pentstemon different from any spoken of before, with an assemblage of smaller plants. I have been continually on my feet, scarcely three nights in one encampment. As I have

accidentally met with a Mr McDonald on his return from a hunting excursion in the south, and the same person I accompanied on a few days' march last August on the Willamette River, he has kindly offered to take the result of my labours for the last month, which I willingly accept. The collection consists of three bundles of dry plants (ninety-seven distinct species), forty-five papers of seeds, three Arctomys, and one curious rat, which I hope you will receive safe. On the evening of the 9th of June, I received a letter from Mr Goode, and one from Mr Booth by an Indian from the coast. Mr McLoughlin did not deem it safe to send what he conceived your letter, and very properly kept it until his people should return from the ocean. I am just going down the river, two days' march, in hopes of receiving yours, for my patience is completely exhausted; and as I am now ready to bend my steps again to the upper country, I cannot well go without hearing from you. I have not a moment to spare, as the people are starting. May I ask Mr Goode to make a note to Mr Atkinson saying I am well, who will inform my brother as I have no time to write him? I can only think of Mr Turner, Lindley, and Munro. It is impossible I can write them a single line. — I am, Dear Sir, your obedient servant, D. Douglas.

To Joseph Sabine, Esq.,

P.S. — July 11, on the Great Falls of the Columbia at sunrise: I arrived here last night and have secured your communication. To say I am happy would only convey but a faint idea of rapture I enjoy in hearing from you. I have no time; am just setting out again for the upper country and am glad to think I may reach the coast before the ship sails for England. — D.D.

Embarked at 10 a.m. and proceeded with a speed not less than twelve miles an hour, the river being at its greatest height. About two a strong wind from the west set in, which greatly retarded our progress and obliged us to put on shore an hour before sundown. As no salmon could be caught, there being too great a swell for the canoes to go out, and having nothing to eat a horse was killed, part of the flesh of which and a mouthful of water I made my supper. Very warm during the day, 97° Fahr. in the shade; thunder and vivid sheet-lightning during the night. Having no tent, was dreadfully annoyed by mosquitoes.

Monday July 10th After a cheerless night and but little refreshed, resumed my route at day break, the morning being

calm and pleasant; put ashore at Day's River, a southern branch
of the Columbia, at eleven o'clock to breakfast, which occupied
an hour. As there was enough boiled last night no fire was
necessary except for smoking. This part of the river being very
rapid, gained already about fifty or fifty-five miles. While eating
my food an Indian who was standing alongside of me managed
to steal my knife which was tied to my jacket by a string, and
being the only one used for all purposes I was loth to part with
it. I offered a reward of a little tobacco for its recovery, without
effect. At last I commenced a search and found it secreted
under the belt of one of the knaves. When detached he claimed
the premium, but as he did not give it on the first application,
I paid him, and paid him well, with my fists that he will, I
daresay, not forget the Man of Grass for some days to come.
As the wind blew up the river very strong after midday, I had
to put to shore twelve miles above the Great Falls, not being
able to get round a rocky island at this part of the river.
Remained until 4 p.m. when the wind abated, and then proceeded. On my gaining the Great Dalles six miles below the
Great Falls at seven o'clock, I observed smoke rising among the
rocks; thinking it to be Indians fishing, I walked to the lower
end in quest of salmon, but instead of Indians I was delighted
beyond measure to find it the camp of the brigade from the
sea. I cannot describe the feeling which seizes me even on
seeing a person again, although I am but partially acquainted
with them. After travelling in the society of savages for days
together and can but speak words of their language, assuredly
the face of a Christian although strange speaks friendship. It
was the more agreeable to me as I am previously acquainted
with all the persons in authority; and I should be destitute of
every feeling of gratitude if I did not mention the kindness and
hospitality shown by all. One caused water to be brought me
to wash, while another was handing me a clean shirt, and a
third employed himself cooking my supper. My old friends Mr
Work and Mr Archb. McDonald handed me my letters, which
were grasped greedily and eagerly broken open. Received one
from Joseph Sabine, Esq., and Mr Munro one, both gratifying;
one from Mr Atkinson, and one from my brother, with a note
from Mr McLoughlin and some other friends at the sea. There
is a sensation felt on receiving news after such a long silence,
and in such a remote corner of the globe more easily felt than
described. I am not ashamed to say that (although it might be
thought weakness by some) I rose from my mat four different

times during the night to read my letters; in fact before morning I might say I had them by heart — my eyes never closed.

Monday 17th July In company with Mr Work and McDonald started on a journey by water with a party of twenty-eight men for the fork of the Snake River, about 150 miles from the Columbia, and as their marches would be short, I hope to put myself in possession of most of the plants found along the banks of that river. Camped fifteen miles up that river.

Tuesday 18th - Monday 24th July 1826 This river is of considerable magnitude, some places 250 to 300 yards broad, in many places very deep and rapid: at least four times as much water as there is in the Thames; the general course east. Twenty-five miles from its junction with the Columbia from an undulating dry, barren country it changes to high rugged mountains, and not a blade of grass to be seen except in the valleys or near the springs, where a little withstands the intense heat. Rose always at daybreak and camped about 10 a.m. and rested till 3 or 4 p.m., the heat being too great for any exertion in the middle of the day, when we made generally fifteen or twenty miles further in the cool of the evening. Some idea may be formed of fatigue experienced on this voyage, when the thermometer frequently stood from 98° to 106° of Fahr, in the shade, destitute of a screen from the scorching sun. The only thing I might say that renders it superior to the deserts of Arabia is abundance of good water enjoyed in inland voyages. That excepted, there is but little difference. Salmon are caught in the river and in some of its branches near the Rocky Mountains, but by no means so plentiful as in the Columbia nor of such good quality. We obtained occasionally a few of them from the Indians, but their extreme indolence prevents them from catching barely what serves themselves. Our general fare was horse-flesh cooked by boiling, and sometimes roasted on the point of a spike before the fire. I learned that the wants of the natives are simple and they require but little to support life in original simplicity. From the oppressive heat I found great relief by bathing morning and evening, and although it causes weakness in some degree, I have some doubt if I had not I should not have been able to continue my trip. On Monday arrived at the branches of the river at dusk, where was a camp of three different nations, upwards of five hundred men able to bear arms. One called Pelusbpa, one the Pierced nose (Chawhaptan) and Chamuiemuch.

The chiefs or principal men came and stayed with us till bed-time and presented some favourite horses.

Tuesday 25th - Monday 31st July As I understood from my companions that their stay would be for a few days, I was desirous of making a trip to the mountains, distant about sixty miles, the same ridge I visited last month further to the south-east; but as they had not yet made any arrangement with the natives, it was not thought prudent to go from the camp, so I was guided by their advice. On Wednesday a conference was held and ended favourably and with great splendour by dancing, singing, haranguing, and smoking. All were dressed in their best garments, and on the whole presented a fine spectacle and certainly a new one to me. On Thursday at daylight — Mr McDonald having been so kind as to send one of his men (Cock de Lard) with me, more as a companion than guide, for he was as much a stranger as myself — we set out in a south-easterly direction, the country undulating and very barren. In the course of the day passed only two springs, and as I was uncertain if more were near and the day far spent, I camped at four o'clock in the afternoon. Found only one species of Pentstemon and a few seeds. Very warm. On Friday reached the mountains at nine o'clock and took my breakfast (dry salmon and water) among some very large trees of *Thuya occidentalis,* the spot pointed out to me by the Indians where Lewis and Clarke built their canoes, on their way to the ocean, twenty-one years ago. I left my men to take care of the horses at the foot of the mountain, while I ascended to see if it afforded anything different from what I had seen before. Reached the highest peak of the first range at 2 p.m. on the top of which is a remarkable spring, a circle 11 feet in diameter, the water rising from 9 inches to 3½ feet above the surface, lowering and rising at intervals, in sudden gushes; the stream that flows from it is 15 feet broad, and 2½ deep, of course running with great force as its fall is 1½ foot in 10 and it disappears at the foot of the hill in a small marsh. I could find no bottom to the spring at the depth of 60 feet. Surrounding the spring there is a thicket of a species of *Ribes* belonging to section *Grossulariae,* 12 to 15 feet high, with fine delicate fruit of a very superior flavour and large, nearly as large as a musket-ball. This fine species I have not seen before; should it prove new, I hope it may be called *R. munroi,* as I have called the spring Munro's Fountain; at the same time how delighted he would feel to see such in the garden. Found in seed, nearly ripe, *Ribes viscosissimum,* and, lest I should not meet with it in a better state, gathered a quantity of it. Found a few seeds of *Paeonia,* but not so ripe as I should have wished, with a small species of *Vaccinium* and a few seeds of *Xylosteum* which I saw in blossom on the mountain near Spokane in May. I joined the man and horses at six o'clock and set out for my encampment of last night. On arriving and looking for something to eat, I found that only salmon for one day had been put in by the man in mistake, and both having a good appetite we mutually agreed to make for the camp. Set out at dusk (Cock de Lard undertaking to be guide) before the moon rose, at least before it became visible; he took us out of the way about ten miles. Arrived at the camp at sunrise, when I threw myself down in one of the tents to sleep. I had not been asleep more than two hours when I was hurriedly aroused to take on myself the profession of a soldier, a misunderstanding having taken place between the interpreter and one of the chiefs; the latter accusing the former of not translating faithfully, words became high till at last the poor man of language had a handful of his long jet hair torn out by the roots. On the Indian being reproved, he went off in a fit of rage and summoned his followers, amounting to seventy-three men. All arrived and came to our camp with their guns cocked and every bow strung. As every one of our party had done all in his power that it should be mutually and amicably adjusted and been refused, every one seemed more careless for the result than another. We (thirty-one of us) stood to our arms and demanded if war was wanted; it was answered 'No, we want only the interpreter killed, and as he was no chief there could be but little ill done'. They were told that whatever person we had in our party, whether chief or not, or if it was only an Indian under our protection, should they attempt to kill or disturb him in the least, certainly they would know we had been already in war. The coolness that seemed to be the prominent feature in our countenance had the desired effect of cooling their desire for war and made them glad to ask for peace, which on our part was as willingly granted them. Many speeches were made on the occasion, and, if it may be allowed to judge from gesture and the language of nature, many of them possess qualifications that would be no disgrace to a modern orator. Although there is much repetition in their harangues, delivered with much vehemence and intense feeling, they are uniformly natural and are certainly calculated either to tie the knot of affection and sympathy, or rouse the mind to discord and war. I have observed speakers and hearers so overcome that they sobbed and cried aloud, and the proceedings delayed until they recovered. This affair was concluded in the usual way — exchange of presents. Although friendship had again been restored, it would have been imprudent to have gone from the camp; therefore I employed myself putting in order those collected and airing some seeds. On Sunday at midday we rose camp and pitched on the northern shore of the north branch.

4

1826
The Snake River and Spokane

The Blue Mountains still intrigued Douglas and he determined to visit them again, and to push up the Snake River to Spokane and Kettle Falls, but first he went down the Columbia to collect his mail.

Monday July 31st - Friday August 4th Early in the morning I had the plants and seeds which were collected carefully secured in one of the saddle-bags. Parted with Mr McDonald, who descended the river; and Mr Work with two men and myself took our departure overland in a north-easterly course to Kettle Falls on the Columbia. On gaining the top of the hills near the river we found the road good in many places and in others very bad, with badger and rabbit holes, which at this season are covered with grass, rendering them more dangerous. Made about forty-five miles. Camped at mid afternoon on a low piece of ground where there was water; passed only one spring in the course of the day. The whole country destitute of timber; light, dry, gravelly soil, with a scanty sward of grass. Gathered seeds of two species of *Astralagus*. Thermometer 97°; heavy dews during the night; minimum heat 53°; rain towards daybreak. Tuesday and Wednesday started early in the morning and went briskly on till eleven, when we halted to breakfast and rest during the heat of day. I opened my saddle-bags and exposed my seeds, lest they should have been wet or damp with the rain last night, and then made a circuit on the high grounds to the south, leaving my watch with Mr Work to know when to start, and as they would pass near my track would pick him up. Remaining longer than expected I went back and found him and the two men fast asleep; packed up again and proceeded till dusk, when we camped under a solitary poplar on the margin of a stagnant pond full of *Nuphar luteum, Advena,* and one species of *Potamogeton*; water very bad. Warm during the day,

F

a fanning wind during the night, which prevented us from being troubled with mosquitoes. Gathered some seeds of *Mimulus albus* (sic) and laid in specimens, and put up a paper of seeds of a species of the same genus. On Wednesday came to the high grounds on the south side of the Spokane River, which is intersected with belts of wood. We intended to stop at a small rivulet in the second one, but our guide (one of the men) missed the road and made too far to the east, where we could get no good water until three o'clock, when we made a stop to take breakfast near a small lake with fine cool water. Made a stay of two hours and proceeded from this lake due north through woods and plains until we came to the banks of the Spokane River, two miles below the falls of that river; the fall is a perpendicular pitch of 10 feet across the whole breadth of the river. Proceeded down the south side of the river seven miles, and as night stole in on us, and having to pass some very high perpendicular rocks, we chose to camp rather than go to the old establishment so late at night. Gathered some seeds of a species of *Lilium* growing in open or partially shady woods. I am sorry that no seeds of *Lilium pudicum* can be found perfect, and, as I could not get any, dug up some roots — perhaps they may keep to England. Warm; thermometer 99° maximum, 56° minimum. Heavy dew.

Thursday and Friday 3rd and 4th August At nine o'clock in the morning crossed the Spokane River to the old establishment on the south side, where we found old Mr Finlay, who gave us abundance of fine fresh salmon from his barrier, placed in a small branch of the main river. After breakfast, and having the horses crossed, left that place at noon for the Columbia. An hour's ride from that place passed the Indian camp on the north side of the river, where they were employed fishing. Their barrier, which is made of willows and placed across the whole channel in an oblique direction, in order that the current which is very rapid will have less effect on it, has a small square of 35 yards enclosed on all sides with funnels of basket-work (just made in the same manner as all traps in England), and placed on the under side, through which the salmon passes and finds himself secure in the barrier. When the spearing commences, the funnels are closed with a little brushwood. Seventeen hundred were taken this day, now two o'clock; how many may be in the snare I know not, but not once out of twelve will they miss bringing a fish to the surface on the barb. The spear is pointed with bone and laced tight to a pointed piece of wood a foot long and at pleasure locks on the staff and comes out of the socket when the fish is struck; it is fastened to the staff by a cord. Fifteen hundred and sometimes two thousand are taken in the course of the day. Camped on my way to Spokane. Gathered a few seeds of a strong species of currant with white blossoms. Cold during the night.

On Friday as usual started at daybreak. About seven o'clock I went off the way to gather some more seeds of the *Ribes* I saw last night. As I found them but scanty it took up a considerable time picking a sufficiency. Overtook my companions ten miles beyond Barrière River, which I had to swim in spring. Rode across it today, halted at noon to breakfast, and proceeded on at two o'clock. On crossing Cedar River, a small rapid stream nine miles from the Columbia, my horse on gaining the opposite shore, which is steep and slippery, threw back his head and struck me in the face and I was plunged head foremost in the river. I fortunately received no further injury than a good ducking and got wet what seeds I had collected during the day, which were in my pockets and knapsack with my note-book. Arrived at Kettle Falls at 7 p.m. having been absent two months, and was cordially and hospitably entertained by Mr Dease.

Wednesday 16th As I learned from Mr McLoughlin that the vessel would not sail for England until the 1st of September, and as the bulk of my plants are at the sea, and having a collection of seeds amounting to 120 of this year's gleanings, and at the same time, as I also learn that this may be the last ship for some years going direct for England, makes me very desirous of sending all that is possible by her. Should Mr Simpson (the Governor) not arrive in a few days, I must endeavour to reach the ocean by some means or other, taking with me my collection of seeds to be sent home. Today packed in a small box five quires of paper, three brown and two cartridge, sixty-nine papers of seeds — a portion of those collected on my journey on the Snake River and since my arrival here — with two linen shirts to be sent across the Rocky Mountains, where I will find it early in June at Fort Edmonton. Very warm. Mr Dease spoke to the Little Wolf, a chief of the Okanagan tribe, about guiding me to that place. The river which is still high with the cascades, rapids, dalles, and whirlpools, renders it impossible to go by water without six or eight men in a canoe; and indeed there are none at this place sufficiently large at present.

August 17th   Packed one bundle of dry plants in my trunk among my little stock of clothing, consisting of one shirt, one pair of stockings and a night-cap, and a pair of old mitts, with an Indian bag of curious workmanship, made of Indian hemp, *Apocynum sp., Helonias tenax,* and eagles' quills, used for carrying roots belonging to the Cootanie tribe, whose lands lay on the shores of the small lake called Cootanie Lake, the source of the Columbia, and that small neck of land at the head-waters of McGillivray's River. An old quarrel of nine years' standing existing between them and the tribes on the Columbia lakes, sixty miles above this place, who are here at present at the salmon fishing at the Falls, gave Mr Dease and every other person much uneasiness. The parties met stark naked in our camp, painted, some red, black, white, and yellow, with their bows strung, and such as had muskets and ammunition were charged. War-caps of calumet-eagle feathers were the only particle of dress they had on. As one was in the act of letting the arrow from his bow, aiming at a chief of the other party, Mr Dease fortunately brought him a blow on the nose which stunned him. The arrow grazed the skin and passed along the rib opposite the heart without doing much injury. The whole day was spent in clamour and haranguing, and as we were not sure what might be the result, we were prepared for the worst. Mr D. proposed that they should make peace tomorrow, and that it would be much better they should go to each other's lands as friends than butchering each other like dogs. His advice they said they should follow; that they would come early in the morning. The Wolf, being one of the principals on one side, told me he cannot go tomorrow, as the peace is to be made, which could not be well done without his presence.

Friday 18th   Bustle and uproar; towards evening peace was signed and sealed by an exchange of presents; and as there is to be a great feast on the occasion, the Wolf is uncertain when he can be spared from his office. As my time is very short, Mr D. spoke to an Indian who is in the habit of attaching himself to the establishment and going on journeys with his people, to go with me — to which he agreed at once. So I will start tomorrow early.

Saturday 19th   Detained till eight o'clock, although Mr Dease had sent out early for the horses, but being far off could not be caught sooner. I had put up for my journey some dried meat of buffalo flesh, a little sugar and tea, and a small tin pot. My gun being left, out of order, in the lower country, Mr Work kindly gave me the loan of a double-barrelled rifle-pistol; and going alone as I am, it is perhaps much better to have as little as possible of tempting articles about me. I left this delightful place highly gratified, having made a tolerable addition to my collection and received every kindness from the hospitable people which they had in their power to show. Being short of clothing, Mr Dease gave me a pair of leather trousers made of deerskin and a few pairs of shoes, which in my present state were very acceptable. He provided me with three of his best horses, one for my guide, one for carrying my little articles, and one for myself. The only thing in the way of clothing except what was on my back, was one shirt and one blanket, and in this shape I set out for Okanagan. It was my intention to have gone by water, but was dissuaded by Mr D., that part of the river at this season being high and from numerous cascades and rapids perhaps dangerous. Proceeded along the south bank of the Columbia, intending to cut the angle between the Columbia and Spokane Rivers. My path very mountainous and rugged, in many places covered with timber of the same sorts as are commonly seen abundantly over all the country. Nothing occurred today, my guide (to whom I cannot speak a single syllable) seems to conduct himself very well. Camped at sundown near a small spring, surrounded by a thicket of birch and willow, at the foot of a high conical hill ten miles from the Spokane River.

Sunday 20th Shortly after two o'clock I had my horses saddled, and, the ground being very uneven and stony, drove them before me; and the moon shining delightfully clear, I found it by far more pleasant travelling during the night than the day. Arrived on the Spokane River nine miles from the Columbia, where there was a large number of Indian lodges, being a fishing ground. After making a short stay and presenting them with a little tobacco, four of them accompanied me two miles further down the river, where they assisted me in crossing the horses and carried myself and all my property across in their canoe. At ten o'clock left the woody country and began my course through a trackless barren plain, not a vestige of green herbage to be seen, soil gravel and sand. About one o'clock I halted, to rest the horses and take some breakfast, opposite the Grand Rapids — having already made nearly fifty miles — and made a small pan of tea, which I let stand till it cooled and settled, and then sucked the water of the leaves. In the interval I gleaned a few seeds, bathed in the rapid, which recruited me greatly, and again in the cool of the evening resumed my route, course west. Towards dusk came to a small pool of stagnant water, very bad, and having nothing to qualify it I was urged to continue till eleven o'clock, when I came to a small spring, but without a single twig for fuel. I made an effort to boil my little pan with dry grass, a large species of *Triticum,* but was unable to succeed. Being an old encampment I fortunately found some horse-droppings, by the aid of which and the grass mentioned I managed to make some tea, when shortly afterwards I laid myself down to sleep on the grass. I have a tent, but generally am so much fatigued that the labour of pitching it is too great. Here it could not be done for want of wood, and tent-poles cannot be carried.

Monday 21st Today I overslept myself; started at four o'clock. The country same as yesterday; at eight passed what is called by the voyageurs the Grand Coulee, a most singular channel and at one time must have been the channel of the Columbia. Some places from eight to nine miles broad; parts perfectly level and places with all the appearances of falls of every extraordinary height and cascades. The perpendicular rocks in the middle, which bear evident vestiges of islands, and those on the sides in many places are 1500 to 1800 feet high. The rock is volcanic and in some places small fragments of vitrified lava are to be seen. As I am situated, I can carry only pieces the size of nuts. The whole chain of this wonderful specimen of Nature is about 200 miles, communicating with the present bed of the Columbia at the Stony Islands, making a circular curve 1¼° further south, and of course longer than the present chain. The same plants peculiar to the rocky shores of the Columbia are to be seen here, and in an intermediate spot near the north side a very large spring is to be seen which forms a small lake. I stayed to refresh the horses, there being a fine thick sward of grass on its banks. The water was very cold, of a bitterish disagreeable taste like sulphur. My horses would not drink it, although they had had no water since last night. At noon continued my route and all along till dusk. The whole country covered with shattered stones, and I would advise those who derive pleasure from macadamised roads to come here, and I pledge myself they will find it done by Nature. Coming to a low gravelly point where there were some small pools of water with its surfaces covered with *Lemna,* or duck weed, and shaded by long grass, one of the horses, eager to obtain water, fell in head foremost. My guide and myself made every effort to extricate it, but were too weak. As I was just putting some powder in the pan of my pistol to put an end to the poor animal's misery, the Indian, having had some skin pulled off his right hand by the cord, through a fit of ill-nature struck the poor creature on the nose a tremendous blow with his foot, on which the horse reared up to defend himself and placed his fore-feet on the bank, which was steep, when the Indian immediately caught him by the bridle and I pricked him in the flank with my pen-knife, and not being accustomed to such treatment, with much exertion he wrestled himself from his supposed grave.

Tuesday 22nd Last night being very warm, with the whole firmament in a blaze of sheet lightning, and parched to a cinder, I passed a few miserable hours of rest but no sleep, and as usual set out before day; and, my road being less mountainous, with little exertion I found myself on the Columbia at midday opposite the establishment. Seeing an old man spearing salmon I had the horses watered and hobbled, and crossed in a small canoe with my guide. Here I found my old friends Messrs McDonald and Ermatinger, who received me with every kindness. After washing and having a clean shirt handed me, I sat down to a comfortable dinner. I was glad to find the small box, which I thought might have been overlooked in one of the portages, brought to this place by the former gentleman and left until I should pass. As my time was of great consequence I communicated to them my wish, and immediately they purchased a small canoe for me, and hired for me two Indians to go with me to the junction of the Snake River. In the meantime I wrote to Mr Dease by my old guide, who behaved himself in every way worthy of trust and is to make a stay of two or three days to rest, and I then put up a few seeds and changed some plants collected on the journey. As I felt somewhat wearied I went early to bed; the doors being left open by reason of the heat, and the windows, which are made of parchments being by no means close, gave the mosquitoes free access, I was under the necessity of abandoning the house at midnight and took myself to a sort of gallery over the door or gate, where I slept soundly. Before leaving this place early in the morning after breakfast, I had a little tea and sugar offered me, which I thankfully accepted, and a small tin pot made in the form of a shaving pot, the only cooking utensil. They regretted the only provision they had that would do for carrying was dried salmon, but as I still had, through the goodness of Mr D., my last host, a little dried meal, in that respect I was not so bad. Left at seven o'clock. In passing a long rapid (two and a half miles) about ten miles below the house, I took the precaution to take out my paper, seeds, and blanket, and was walking along the shore with them while the Indians ran the canoe down. When in the middle of the rapid a heavy surge broke over them and swept every article out of it except the dry meat, which being weighty by chance was wedged in the canoe, it being very narrow. The loss of the tea and sugar with the pot was a great one in my present situation, but I considered myself happy, having saved my papers and seeds. As I have said something of the river on my ascent, I need only observe my encampments. Camped at the south of the Piskahoas River at dusk. Today, although I paddled all day — at least served as steersman — I did not feel fatigued but my hands were much blistered.

Thursday 24th By eight o'clock gained the Stony Islands, an extremely dangerous part of the river where the channels are very narrow, not more than 20 to 30 feet broad. As my guides were little acquainted with this part of the river, I hired an Indian of the place to pilot my canoe and after landing her safe below, I paid with a few crumbs of tobacco and a smoke from my own pipe. As I had nothing to cook I ate some crumbs of dried meat and salmon, and when I wanted to smoke kindled my pipe with my lens, so I was not under the necessity of making a stay to kindle a fire. Reached the top of the Priest Rapids at six o'clock, and although late I undertook to run the canoe down, making my old guide (they were father and son) carry my little parcels, he being tired. Night stole in on me too soon, and I was obliged to camp on the north side of the river in the middle of the rapid; four and a half miles.

Friday 25th I could not leave my encampment before daylight, having still four and a half miles of very bad weather. I had left by land an hour before the canoe, and, after waiting nearly an hour at the foot of the rapid, as my guides did not make their appearance I became alarmed for their safety and returned, when I discovered them about a mile and a half above where I halted, comfortably seated in a small cove treating some of

In March and April, from Fort Vancouver to the Kettle Falls   638

In May, Journey to Spokane   150

In June, from Kettle Falls to the Junction of Lewis & Clark's River   414

In June, Journey to the Blue Mountains   190

in July, a Second to the same   137

In July, descending Lewis and Clark's River to its N. and S. branch   140

a 3d Journey to the Blue Mountains from that place   103

From Lewis and Clark's River to Spokane   165

From Spokane to Kettle Falls   75

In Augt, from Kettle Falls to Oakanagan by land   130

From Oakanagan to Fort Vancouver   490

In Sept, Octr, and Novr from the Columbia to the Umpqua River and the country contiguous thereto   593

To the ocean and the Bays North of the Columbia & Islands   125

Daily allowance from my places of rendesvous   000

     3,969

1827

In March and April, the whole chain of the Columbia from the Sea to the Rocky Mountains   995

     7,166

My notes will show by what means it was done —

28 Saturday — last night cold: Min Heat 18°, Mean 30°, obliged to rise down the mill to make fire ☉ Delayed commencing my journey, Mr. — being employed laying the boat and other articles en cache until 8 am when we breakfasted and took leave of the main body of the Columbia at a slow ½ course ☉ passed a low front & wood of a mile and entered a swamp about 3 long, frequently sinking to the knees in water, which was doubly fatiguing from the thin ice on its surface, too weak to bear us up — before a deep muddy brook and entered a second front of wood of an uneven hilly surface ☉ at 11 obliged to have recourse to my Bear's paws or snow shoes, the snow being then deep by the sun's influence 4 to 7 feet deeper — Much annoyed throughout the day, their lacing or netting becoming slack by the wet, and being little skilled in the use of them now and then I was falling head over heels, sinking one leg, stumbling with the other ☉ They sometimes turning

[...] with me. At midday I reached my long wished tree, and lost no time in examining and endeavouring to collect specimens and seeds. New or strange things seldom fail to make great impressions, and often at time are liable to overrate them; and now, lest I should never see my friends to tell them verbally of this most beautiful and *most* immensely large tree, I now state the dimensions of the largest one I could find that was blown down by the wind:— 3 feet from the ground, 57 feet 9 inches in circumference;— 134 feet from the ground, 17 feet 5 inches;— extreme length 215 feet. The trees are remarkably straight— bark uncommonly smooth for such large timber, of a whitish or light brown color; and yields a great quantity of gum of a bright amber color. The large trees are destitute of branches, generally for $\frac{2}{3}$ the length of the tree;— 215 the branches, pendulous, something like the

and the cones hanging from their points like small sugar-loaves in a grocer's shop. It being only on the very largest trees that cones are seen, and the putting myself in possession of 3 cones (all I could) nearly brought my life to an end. Being unable to climb or hew down any, I took my gun and was busy clipping them from the branches with ball when eight Indians came at the report of my gun. They were all painted with red earth, armed with bows, arrows, spears of bone, and flint knives, and seemed to me any thing but friendly. I endeavoured to explain to them what I wanted and they seemed satisfied and sat down to smoke, but had no sooner done than I perceived one string his bow and another sharpen his flint knife with a pair of wooden pincers and hang it on the wrist of the right hand, which gave me ample testimony of their intentions.

The discovery of the Sugar Pine. Photograph of Douglas' Journal, October 25, 1826. *Courtesy Royal Horticultural Society*

Sugar Pine. *Photo Ken Taylor A.R.P.S.*

Douglas Fir at Drumlanrig, Dumfriesshire. This tree was sent by Douglas to his brother who was Clerk of Works at the Castle about 1829, and planted out about 1832. Girth nearly 17 feet in 1978. *Photo J. Davies. Courtesy Duke of Buccleuch*

Cone of Abies Procera, Noble Fir. "I spent three weeks in a forest composed of this tree and day by day could not cease to admire it . . ." *Photo Ken Taylor, A.R.P.S.*

Sir William Hooker, F.R.S. *Gambarella. Courtesy Linnean Society*

Joseph Sabine, F.R.S., Secretary of the Horticultural Society 1810-1830. *W. Read. Courtesy Linnean Society*

McGillivray's Rock, taken from the Canoe Pass. Douglas noted this mountain in May 1827. Whirlpool river and Committee's Punchbowl out of sight in middle distance. *R. M. Patterson 1955*

In the Athabasca Pass, Mt. Evans and Mt. Kane. The Whirlpool river is on the left. R. M. Patterson, 1955

The Committee's Punchbowl. "The small lake at the watershed of the
MacKenzie in Columbia"

Edward Ermatinger. Douglas'
guide through the Athabasca
Pass in 1827. *Public Archives
of Canada*

Fort Edmonton in Douglas' time. *Courtesy Glenbow Institute*

Fort Vancouver, B.C., 1844-46. *From sketch by Captain Warre. Provincial Archives, Victoria, B.C.*

John McLoughlin, M.D., welcoming American settlers to Fort Vancouver in 1834. *Charles Comfort. Courtesy Hudson's Bay Company*

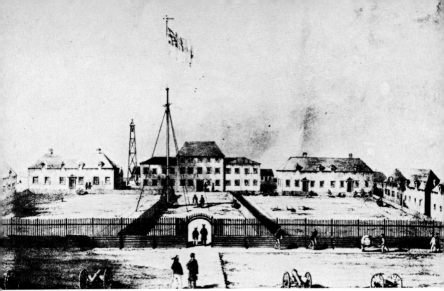

York factory in its heyday in 1853. *Courtesy Hudson's Bay Company*

Norway House. *An early watercolour. Courtesy Glenbow Institute*

Rear view of Jasper House, Rocky Mountains, Jan. 7, 1872. *Provincial Archives, Victoria, B.C.*

Jasper House, Alberta. *From a watercolour by W. G. K. Hinr. Provincial Archives, Victoria, B.C.*

Fort St. James, British Columbia, the most northerly place Douglas reached. *Courtesy Dept. of Travel Industry, B.C.*

Stone Indian Boys. *Public Archives of Canada*

Buffalo hunting on tne Western Prairies. *Courtesy Glenbow Institute*

Portrait of a Stone Indian, 1824. *Sir George Back. Courtesy Glenbow Institute*

Sir George Simpson, Governor
of the Hudson's Bay Company
Courtesy Hudson's Bay Company

John McLoughlin (M.D.). *Provincial Archives, Victoria B.C.*

Sir John Franklin, 1828. *T. Phillips, National Portrait Gallery, London*

rtage on the Hoarfrost
ver. *Sir George Back, 1833.*
urtesy Glenbow Institute

The Cumberland Station of the Church Missionary Society, North West America. *Provincial Archives, Victoria, B.C.*

Franklin's Grouse. The specimen shot by Ermatinger on 1st May 1827 and preserved by Douglas. *Courtesy National Museum of Scotland*

Lupinus Latifolia. One of Douglas' actual specimens. *Courtesy Royal Botanic Gardens, Edinburgh*

their friends to a smoke with some tobacco I had given them the preceding evening. As I had now a fine sheet of water without any rapids, but a very powerful current, I went rapidly on, like the day before scarcely out of my canoe, and arrived at Wallawallah, the establishment near the Snake River, at sundown. I felt so much reduced that I was too weak to eat, and after informing Mr Black of my going to the sea and asking him to procure me two guides to carry me to the Great Falls in the morning, I laid myself down to rest on a heap of firewood, to be free from mosquitoes.

Saturday 26th Wrote to Mr McDonald by the old guide and gave him ten charges of ammunition and a little tobacco to buy his food on the way home, and after obtaining a larger canoe from Mr Black in lieu of my present one, and two guides, I took my leave for the sea at six o'clock. At the foot of a rapid twenty-five miles below I purchased a fresh salmon, the half of which I roasted on a stick for breakfast and reserved the other half for next day, lest I should not get anything. As I knew all the bad places of the river, I went on all night drifting before the stream, taking the steering in turn, and as I had to pass a camp of Indians who are noted pillagers, made me anxious to pass during the night, which I accomplished.

Sunday 27th Precisely at noon I reached the Great Falls, and finding my canoe too heavy to carry over the rocks I left it and hired one to carry me to the Dalles six miles below. Here I purchased a pair of horns of a male grey sheep of the voyageurs, for which I paid three balls and powder to fire them. The Indian had the skin dressed, forming a sort of shirt, but refused it me unless I should give him mine in return, which at present I cannot spare. On the Dalles were at least five hundred to seven hundred persons. I learned that the chief Pawquanawaka, who would have been my last guide to the sea, was not at home; but as I am now in my own province again, and understand the language tolerably well, I had no difficulty in procuring two, and was glad to find one who was well known to me. While he and his companion brought the canoe down the Dalles, after being refreshed with a few nuts and whortleberries I proceeded over a point of land fifteen miles, taking an Indian to assist me in carrying my things. The canoe did not appear till an hour after dark. In the evening a large party of seventy-three men came to smoke with me, and all seemed to behave decently till I discovered that my tobacco-box was off. I had hung up my jacket and vest to dry, being drenched in the canoe descending the Dalles. As soon as I discovered this I perched myself on a rock, and in their own tongue I gave them a furious reprimand, calling them all the low names used to each other among themselves. I told them they saw me only one blanket man but I was more than that, I was the grass man, and was not afraid. I could not recover it. After all the quarrel, I slept here unmolested.

Monday 28th Detained by a strong west wind till eight o'clock, when it became more moderate and I proceeded. Made but little progress; Camped fifteen miles above the Grand Rapids.

Tuesday 29th As the wind increased with the day, I could not venture out in the stream, and even near the shore the waves were so high that I had to carry my property on my back along the high shelving rocks, leaving the Indians to bring down the canoe. Arrived at the village on the Grand Rapids at three and repaired to the house of Chumtalia, the chief, and my old guide last year, where I had some salmon and whortleberries laid before me on a mat. I made a hearty meal and then spoke of procuring a large canoe and Indians to take me to the sea. He offered to go himself, but as he was busily employed curing salmon I was loth to accept his services, and took in preference his brother and nephew, with a fine large canoe, and proceeded down the lower end of the rapid in the evening. Camped on a low bank of sand in the channel where there was no herbage, so of course was not annoyed with insects. Long before daylight I was under way lest I should be detained with wind, which for the last three days rose with the sun. Passed Point Vancouver at sunrise. I had the gratification of landing safe at Fort Vancouver at midday, after traversing nearly eight hundred miles of the Columbia Valley in twelve days and unattended by a single person, my Indian guides excepted. My old friends here gave every attention a wayworn wanderer is entitled to. On their discovering me plodding up the low plain from the river to the house alone, unpleasant thoughts struck them. As the river was never seen higher than it has been this year, and of course caution is requisite in descending, they apprehended I was the only survivor. I confess astonishment came over me to meet a people from whom I had had more kindness a thousand times than I could ever have expected, look so strange on me; but as soon as I dispelled the cloud of melancholy that sat on every brow I had that unaffected welcome so characteristic among people so far from home. I had a shirt, a pair of leather trousers, an old straw hat, neither shoe nor stocking nor handkerchief of any description, and perhaps from my careworn visage had some appearance of escaping from the gates of death. In the list of the little society I have here is a Peter Skeene Ogden, Esq. (brother of the Solicitor General for the Canadas), a man of much information and seemingly a very friendly-disposed person; also Capt. Davidson, of the ship Dryade. I am glad to find that all my collections arrived safe here, except a male curlew and a female partridge. As I had no time to lose (tomorrow being the day appointed for the clearing of the ship), I unpacked and repacked seeds brought with me and those sent before from the interior.

5

1826
The Sugar Pine Saga

The knowledge that there was a massive species of pine growing somewhere in the headwaters of the Willamette River had intrigued Douglas all the summer. In August 1825 he had found some huge pine seeds in the tobacco and shot pouches of Indians and in February the hunter Jean Baptist McKay had procured for him a cone which was over 16 inches long. He managed to persuade Mr McLoughlin to let him go into this relatively unexplored country to seek his "much wished for pine", *Pinus lambertiana*. This hitherto undiscovered species is the largest of all the 96 different pines known to science.

September 14th 1826 On Thursday, I consulted with Mr McLoughlin about my proposed journey to that country south of the Columbia and the Willamette and towards the Umpqua River. I had spoken to Jean Baptist McKay, one of the hunters, who goes sometimes through the country contiguous to the latter river; but as he had been here in July, fully six weeks sooner than expectation, and was gone before I arrived, perhaps I might have difficulty in overtaking him. Mr McLoughlin informed me that a party in a few days would be despatched for that quarter under the superintendence of A. R. McLeod, Esq., a gentleman who has given me much civility, and that there would be nothing to prevent my accompanying him. I could not allow this favourable opportunity to escape, and expressed my thanks to the person in confidence for their assistance. Mr McLeod left on Friday to go by land to Mackay's abandoned establishment on the Willamette, fifty-six miles from its junction with the Columbia, where he is to remain until joined by the remainder of the party, which will leave in a few days.

Saturday 16th - Tuesday 19th Employed making preparations for my march. As my gun has entirely failed me, I am under the necessity of purchasing a new one, which only costs £2. Being a new country and no knowledge whatever south of the Umpqua, each has to confine himself to as little encumbrance as possible; and as nearly the whole must be land carriage, this increases the difficulty. Packed six quires of paper and other little articles for my business, and provided myself with a small copper kettle and a few trifles, with a little tobacco for presents and to pay my way on my return. Of personal property (except what will be on me), one strong linen and one flannel shirt; and as heavy rains may be expected, being near the coast, I will indulge myself with two blankets and my tent. On Monday Mr McLoughlin kindly sent by land, to await me on the Willamette, one of his finest and most powerful horses for carrying my baggage or riding, as he may be required, which is of great service.

Wednesday 20th At midday left Fort Vancouver in company with Mr Manson, one of the persons in authority, and a party of twelve men and one boat., with their hunting implements. Camped five miles up the Willamette, on the coast bank of the river, at dusk.

Thursday 21st Proceeded on our journey at sunrise and camped on the rocks on the Falls, among some drift-wood. Warm during the day, with heavy dews at night.

Friday 22nd The boat being injured in hauling it up the rocks, two hours were spent in gumming; gathered a few seeds of different plants, among which *Pentstemon richardsonii,* perfectly ripe. Proceeded at seven o'clock and put ashore seven miles above to breakfast, which occupied an hour. Reached Mr McLeod's encampment at four o'clock, and after having our tents pitched sat down to supper at dusk. Pleasant, with a beautiful sky in the evening.

September 23rd - 25th Detained longer at our encampment than expectation in consequence of several horses missing, having strayed in the woods.

*September 27t*h Morning spent collecting the horses, and before all were ready for starting it was near noon. Took our course due west, towards the coast; passed a small creek, two and a half

miles on, and camped shortly after two o'clock. Country undulating; soil rich, light, with beautiful solitary oaks and pines
interspersed through it, and must have a fine effect, but being
all burned and not a single blade of grass except on the margins
of rivulets to be seen. This obliged us to camp earlier than we
would have otherwise done. As we had no fresh meat, each
took his piece and went out. I raised two small deer and fired at
one without effect. Mr McLeod is not yet arrived at the camp
and most probably will bring something home. Marched today
five miles.

Thursday 28th Mr McLeod returned shortly after dusk last
night and brought with him one of the Indian guides from the
coast south of the country inhabited by the Killimuks. All
unfortunate in the chase, and although nine small deer were
seen in a group, yet by their keeping in the thickets near the
small stream a few miles from our encampment, prevented the
hunter from approaching them. Started at eight o'clock, keeping
a south-west course. Passed two small streams. About noon, in
a small hummock of hazel and bracken, started four deer,
one of which was killed by one of the hunters with his
rifle at two hundred yards distance. The ball entered the left
shoulder and passed through the neck on the opposite side, yet
she ran three hundred yards before she fell. Camped on the
south side of Yamhill River, a small stream about twenty-five
yards wide; channel for the greater part mud and sand.

Saturday 30th September Cloudy until noon, after-part of the
day clear and fine with a fanning westerly wind. In the morning
dried some of my things which got wet the preceding day.
Started at nine and continued our route in a southerly direction,
on the opposite side of the hill from where we were yesterday.
Most parts of the country burned; only on little patches in the
valleys and on the flats near the low hills that verdure is to be
seen. Some of the natives tell me it is done for the puropse of
urging the deer to frequent certain parts, to feed, which they
leave unburned, and of course they are easily killed. Others
say that it is done in order that they might the better find wild
honey and grasshoppers, which both serve as articles of winter
food. I walked along the low hills but found nothing different.
Saw four large bucks. Reached the camp at four o'clock, cleaned
my gun, had supper and went out with a hunting party. One
female and her kid were killed by the Indian hunter. Mr McLeod

passed a ball through the right shoulder of one and afterwards had two other shots fired with effect, but not mortal. Notwithstanding, although to appearance seriously injured, she made her escape. Returned shortly after dark.

Sunday 1st October Camped at four on the banks of a small stream which falls into the Willamette three miles to the east. In the like journeys Sunday is known only by the people changing their linen, and such of them as can read in the evenings peruse religious tracts, &c., whose tenets are agreeable to the Church of Rome. In the dusk I walked out with my gun. I had not gone more than half a mile from the camp when I observed a very large wasp-nest, which had been attached to a tree, lying on the plain where the ground was perfectly bare and the herbage burned, taken there by the bears. At the same time John Kennedy, one of the hunters, was out after deer and saw a very large male grizzly bear enter a small hummock of low brushwood two hundred yards from me. Being too dark, we thought it prudent to leave him unmolested; perhaps we may stand a chance of seeing him in our way. Marched eighteen miles.

Monday 2nd Early in the morning a small doe was killed near our encampment and was placed on one of the horses. As we passed on the track, I went in search of the large bear seen last night but could see nothing of him. At noon passed two deep gullies which gave much trouble, the banks being thickly covered with brushwood, willow, dogwood, and low alder. Course nearly due south, inclining to the west. As no place could be found suitable for fodder for the horses, we had to travel till four o'clock, when we camped at a low point of land near a woody rivulet. Marched twenty-one miles. My feet tonight are very painful and my toes cut with the burned stumps of a strong species of *Arundo* and *Spiraea tomentosa*.

Thursday 5th After a scanty breakfast proceeded at nine o'clock in a south course. Country more hilly. At one o'clock passed on the left, about twenty-five or thirty miles distant, Mount Jefferson, of Lewis and Clarke, covered with snow as low down as the summit of the lower mountains by which it is surrounded. About twenty miles to the east of it, two mountains of greater altitude are to be seen, also covered with snow, in an unknown tract of country called by the natives who inhabit it 'Clamite'.

On the low hills observed *Pinus resinosa* of very large dimensions, 4 to 6 feet in diameter, 90 to 130 feet long. Cones not perfectly ripe. (Secure specimens on your return). Killed a very large grey squirrel, 2 feet long from the point of the tail to the snout. Saw a curious variety of the ground or striped, and also the flying, but could secure neither. Camped on the side of a low woody stream in the centre of a small plain — which, like the whole of the country I have passed through, is burned. One of the hunters killed a small doe shortly after leaving our encampment, which will provide an unexpected supper. Marched nineteen miles.

Friday 6th   At noon we were joined by Jean Baptist Mackay and two Iroquois hunters on their way to Umpqua River. Mackay informs me he had five days ago sent one of his people on to collect the cones I spoke of to him last spring, lest the season should be past before he would be there, he not knowing of our projected journey. The day being very warm and the horses much fatigued, we were obliged to camp earlier than usual, lest we should not get in time to a feeding place. Marched sixteen miles. As I walked nearly the whole of the last three days, my feet are very sore from the burned stumps of the low brushwood and strong grasses.

Saturday 7th   I bargained with Baptist Mackay for a skin of a very large female grizzly bear which he had killed seven days before. I gave him an old small blanket and a little tobacco. This was to make myself an under-robe to lie on, as I found it cold, from the dew lying on the grass. Mackay is to endeavour to kill me male and female, so that I might have it in my power to measure them if not to skin them. John Kennedy had this morning gone out hunting two hours before day, and about ten o'clock was attacked by a large male grizzly bear. He was within a few yards of him before he was discovered, and as he saw that it was impossible to outrun him he fired his rifle without effect and instantly sprang up a small oak-tree which happened to be near him. The bear caught with one paw under the right arm and the left on his back. Very fortunately his clothing was not strong, or he must have perished. His blanket, coat, and trousers were almost torn to pieces. This species of bear cannot climb trees. A party went out in search of him but could not fall in with him.

Monday 9th Morning cloudy; drizzly rain. As we expected wet weather, did not start till noon. Hunters out and fell in with a small herd of elk; but being in the close and almost impenetrable thickets, only one could be secured, which fell receiving eleven shots. At this season the males are very lean and tough eating; weighed about 500 lb. Horns very large, 33 inches between the tips, with five prongs on each, all inclining forward; the two largest, 11 inches long running parallel with the nose and reaching nearly to the nostrils; body of a uniform brown, with a black mane 4 inches long. I am pretty certain this is the sort of animal which I have seen at the Duke of Devonshire's, and unquestionably a very distinct species from the European stag.

Tuesday 10th Morning cool and pleasant, calm and cloudy. Having to cross the same hill I was on yesterday a few miles to the south, and apprehending rainy weather, and the horses poorly off for fodder and ourselves on bad hunting ground, we started at eight o'clock and reached the other side at two p.m. where we encamped in a small woody valley. Besides the conifers usually seen, I observed one not unlike the black spruce, very large but without cones. *P. menziesii, A. balsamea,* abound on the lower parts; *P. resinosa* on the more elevated, all of extremely large dimensions, some 200 to 250 feet high, 30 to 55 feet in circumference.

Friday 13th After securing my little gleanings I left at seven o'clock, and at nine came to a small rivulet that takes its rise in the mountains to the east and discharges itself into the Umpqua River, south of this about thirty-five or forty miles. Here I found Baptist Mackay, who had gone before yesterday for the purpose of hunting, where Mr McLeod joined me shortly afterwards. They brought two small deer, a male and female (entire), so I had an opportunity of measuring them. After resting a short time, Mackay made us some fine steaks, and roasted a shoulder of the doe for breakfast, with an infusion of *Mentha borealis* sweetened with a small portion of sugar. The meal laid on the clean mossy foliage of *Gaultheria shallon* in lieu of a plate and our tea in a large wooden dish hewn out of the solid, and supping it with spoons made from the horns of the mountain sheep or Mouton Gris of the voyageurs. A stranger can hardly imagine the hospitality and kindness shown among these people. If they have a hut, or failing that, if the day is

G

wet, one of brushwood is made for you, and whatever they have
in the way of food you are unceremoniously and seemingly with
much good-will invited to partake. After smoking with a few
straggling Indians belonging to the Umpqua tribe, we resumed
our route on the banks of the small stream; track mountainous
and rugged, thickly covered with wood in many places; and in
some parts, where *Acer circinnatum* forms the under-wood, a
small hatchet or large knife like a hedge-bill is indispensably
necessary. Camped at three o'clock five miles further down the
stream, having gained this day eleven miles. Mackay's hunter
returned this evening from the Umpqua River, but in consequence of some misunderstanding he only brought a few seeds
baked on the embers of the pine. As there are in the party two
individuals of that nation who both talk the Chenook tongue
fluently, in which I make myself well understood, from the
questions I have put to them and the answers given, I am almost
certain of finding it in abundance. Should I fail, I shall make
my way through that very partially known country called
'Clamite', to the north-east of where I am now, where I will
find it without any doubt according to Mr McDonald, who passed
there in September 1825, and this person being present when
it came first under my notice, I requested he would look for
it in the Cascade range of mountains or through that mountainous country between Mount Hood and another high snowy
mountain to the south of it, which I have denominated Mount
Vancouver. He found it there, being then going on a year's
journey; not knowing where to, and no doubt entertaining views
of no very pleasant nature, he of course could not bring me
any. Failing my present trip, I shall return in that direction and
will probably come on the Columbia at the Great Falls.

Monday 16th At nine o'clock began our day's march, in nearly
a westerly course. Passed two miles of open hilly country, intersected by several small streams, where we entered the thick
woods. Passed three ridges of mountains, the highest about 2700
feet. Mr McLeod and I took the lead and were followed by
Baptist Mackay and two hunters, hewing the branches down
that obstructed the horses from passing. The whole distance
not so much as a hundred yards of ground on the same level,
and the numerous fallen trees, some of which measured 240
feet long and 8 feet in diameter — I am aware that it could
hardly be credited to what a prodigious size they attain. The
rain of the two days before rendered the footing for the poor

horses very bad; several fell and rolled on the hills and were
arrested by trees, stumps and brushwood. As I apprehended some
accident, I thought it prudent to carry my gleanings on my
back, which were tied up in a bear's skin. Found one shed horn
of Black-tailed deer: the temptation was too great, so I tied
it on the bear-skin bundle. Appears to be a larger animal than
the Long-tailed Deer. Every two or three hundred yards called
for a rest, and two or three times in the course of the day
for a pipe of tobacco, and in this way did I drag over the most
laborious and tedious marches I have experienced for many
days. On reaching the summit of the last hill the desired sight
of the Umpqua River presented itself to our view, flowing
through a variable and highly decorated country — mountains,
woods, and plains. Half an hour's walk carried us to the banks,
where we encamped on the east side in the angle where the
little river we marched along today falls into it. Arrived at
five o'clock having marched seventeen miles. Having scarcely
any fresh food, Mr McLeod and McKay went out in hopes of
killing deer, while I employed myself chopping wood, kindling
the fire, and forming the encampment; and after, in the twilight,
bathed in the river: course north-west; bed sandstone; ninety
yards broad; not deep, but full of holes and deep chinks worn out
by the water. Two hundred yards below is a small rapid in
several channels and small grassy islands; will never admit of
any barge larger than a ship's jolly-boat, by the numerous rocks
that rise above water and the rapids. The distance from the
ocean cannot exceed thirty or thirty-five miles, as I observe
Menziesia ferruginea and *Tsuga,* which keep along the skirts
of the ocean. Only a few of the horses now came (eight
o'clock) and although the moon is shining, they will not come
till tomorrow. Mine is among the number, so that I have nothing
to lie down on although very tired. Mr McLeod returned unsuccessful, having wounded a large deer which perhaps will be
found in the morning. Evening fine.

Tuesday 17th Last night I sat up by the fire until ten o'clock,
when we learned that some of the people camped in the valley
over the first mountain and some only half-way there, the horses
being worn out. Mr McLeod very kindly gave me his own
blanket and buffalo-robe, and reserved for himself two great-
coats, as the horse with my articles was one of the furthest
behind. Morning dull and heavy, noon fine and clear, wind
easterly. Having nothing for breakfast, Mr McLeod and Mckay

and myself went out on the chase. The deer wounded last night
by Mr McLeod was found, the ball having passed through both
shoulders and a second was still necessary before she could
be taken. Mackay made a fine shot at the distance of two
hundred yards, his ball passing through the chest, upon which the
deer took to the water and was swimming to the opposite side
when he passed a second in at one ear and out of the other.
Arrived at the camp at twelve o'clock in fine spirits. The deer
floated down the stream and was dragged to shore by an Indian
boy. After a comfortable meal between one and two o'clock, I
turned my specimens and exposed some of the last gathered seeds
and cones of *P. menziesii* and then made a turn down the
river, but found nothing different from yesterday. The horses
with my articles arrived at four o'clock in a sad condition. The
tin-box containing my note-book and small papers broken and
the sides pressed close together; a small canister of preserving
powder in a worse state; and the only shirt except the one now
on my back worn by rubbing between them like a piece of
surgeon's lint. In the evening, arranged my papers and found
nothing materially injured. I am glad I took the precaution of
carrying the specimens of seeds and plants on my back, otherwise they would have been much destroyed. The country towards
the upper part of the river appears to be more raised and
mountainous, and perhaps will afford my wished-for pine being
nearer the spot described to me in August 1825 by an Indian
while on the Willamette, in whose smoking-pouch I found some
of its large seeds. Should the morning be fine and any provision
killed to take with me, I intend to start for a few days. Baptist
Mackay has given me one of his Indian hunters, a young man
about eighteen years old, as a guide; of what nation he belongs
to he does not know, but tells me he was brought from the
south by a war party when a child and kept as a slave until
Mackay took him: he is very fond of this sort of life and has no
wish to returning to his Indian relations. He speaks a few words
of the Umpqua tongue and understands the Chenook, so I will
have no difficulty in conversing with this, my only companion.
Keep the horns of this large deer, which I will measure at a
more convenient season. Evening fine.

One can surmise that they made a base camp at or near
Elkton and it took several days before all the stragglers and
horses came in. Baptist Mackay lent Douglas an eighteen year
old Indian hunter and on 18th October they set off. Unfortunately

Douglas had an accident and had to return, and it was not until October 23rd that he was fit enough to start again. McLeod and Mackay also set off, but westwards down the Umpqua to the Pacific to explore the coast line.

Monday 23rd Morning cloudy and calm. Last night after supper Mr McLeod kindly spoke to the chief for his son to accompany me to the upper country while he went with himself along the coast, to which he agreed. The road being very hilly, woody, and difficult to pass over, I did not think it necessary to accept of any more horses than what would carry my blanket and paper, which were two — one for my guide and one for the articles. Started at ten o'clock and passed along the same side of the river and crossed at the chief's lodge where I was some days ago. They readily carried me across in the canoe and behaved very civilly. Made a short stay and crossed over the same point of land to the place where I attempted making a raft and without success, where I killed a small doe which gave me a little hope at the beginning of my march. My guide, by kindling a small fire, brought two men and a canoe from their lodge two miles above, round a thick woody point, who instantly took me across and guided the horses to a shallow part of the river where they forded it and received no injury. Proceeded on the opposite side and camped a short distance from the lodge. I could not utter a single syllable, but by signs they kindled my fire, brought me water, nuts, roots of *Phalangium quamash,* and the sort of meat made of the *Syngenesious* plant spoken of before and some salmon-trout. Finding them not only hospitable but kind in the extreme, I gave them all the flesh of the deer except one shoulder, some presents of beads, rings, and tobacco.

Tuesday 24th Morning cloudy, raw, and dull. My new friends had during the night gone to a small rapid a mile below for the purpose of spearing trout for me and awoke me this morning long ere day to eat. Left my camp at daylight, and passed a low level rich plain four miles long, along the banks of the river, where I entered a thick wood five miles broad and came again on a bend of the river, where I stayed a short time to refresh my horses, being noon; and although having only made nine miles they were much fatigued by the last five being through deep gullies, rocky and obstructed by fallen timber. About two o'clock resumed my course due east over a bare hill 3000 feet above the level of the river, and on gaining the other side crossed a small stream twenty-five or thirty yards broad, shallow but rapid, where I entered a second point of wood three miles broad, hilly, and on almost impenetrable thicket. On leaving it about five o'clock, I was urged to creep along for a feeding place for my horses, which I found a mile and a half further on, and before I had made my encampment the rain was falling in torrents. Cooked the last of my deer flesh and boiled a few ounces of rice for supper, and lest I should not see any Indians, I can only afford one meal a day. Marched seventeen miles.

Wednesday 25th Last night was one of the most dreadful I ever witnessed. The rain, driven by the violence of the wind, rendered it impossible for me to keep any fire, and to add misery to my affliction my tent was blown down at midnight, when I lay among bracken rolled in my wet blanket and tent till morning. Sleep of course was not to be had, every ten or fifteen minutes immense trees falling producing a crash as if the earth was cleaving asunder, which with the thunder peal before the echo of the former died away, and the lightning in zigzag and forked flashes, had on my mind a sensation more than I can ever give vent to; and more so, when I think of the place and my circumstances. My poor horses were unable to endure the violence of the storm without craving of me protection, which they did by hanging their heads over me and neighing. Towards day it moderated and before sunrise clear, but very cold. I could not stir before making a fire and drying part of my clothing, everything being completely drenched, and indulging myself with a fume of tobacco being the only thing I could afford. Started at ten o'clock, still shivering with cold although I rubbed myself with my handkerchief before the fire until I was no longer able to endure the pain. Went through an open hilly country some thirteen miles, where I crossed the river to the south side near three lodges of Indians, who gave me some salmon such as is caught in the Columbia and at this season scarcely eatable, but I was thankful to obtain it. Made a short stay and took my course southerly towards a ridge of mountains, where I hope to find my pine. The night being dry I camped early in the afternoon, in order to dry the remaining part of my clothing. Travelled eighteen miles.

Thursday 26th Weather dull and cloudy. When my people in England are made acquainted with my travels, they may perhaps think I have told them nothing but my miseries. That may be very correct, but I now know that such objects as I am in quest of are not obtained without a share of labour, anxiety of mind, and sometimes risk of personal safety. I left my camp this morning at daylight on an excursion, leaving my guide to take care of the camp and horses until my return in the evening, when I found everything as I wished; in the interval he had dried my wet paper as I desired him. About an hour's walk from my camp I was met by an Indian, who on discovering me strung his bow and placed on his left arm a sleeve of racoon-skin and stood ready on the defence. As I was well convinced this was prompted through fear, he never before having seen such a being, I laid my gun at my feet on the ground and waved my hand for him to come to me, which he did with great caution. I made him place his bow and quiver beside my gun, and then struck a light and gave him to smoke and a few beads. With my pencil I made a rough sketch of the cone and pine I wanted and showed him it, when he instantly pointed to the hills about fifteen or twenty miles to the south. As I wanted to go in that direction, he seemingly with much good-will went with me. At midday I reached my long-wished *Pinus* (called by the Umpqua tribe *Natele*), and lost no time in examining and endeavouring to collect specimens and seeds. New or strange things seldom fail to make great impressions, and often at first we are liable to over-rate them; and lest I should never see my friends to tell them verbally of this most beautiful and immensely large tree, I now state the dimensions of the largest one I could find that was blown down by the wind: Three feet from the ground, 57 feet 9 inches in circumference; 134 feet from the ground, 17 feet 5 inches; extreme length, 215 feet. The trees are remarkably straight; bark uncommonly smooth for such large timber, of a whitish or light brown colour; and yields a great quantity of gum of a bright amber colour. The large trees are destitute of branches, generally for two-thirds the length of the tree; branches pendulous, and the cones hanging from their points like small sugar-loaves in a grocer's shop, it being only on the very largest trees that cones are seen, and the putting myself in possession of three cones (all I could) nearly brought my life to an end. Being unable to climb or hew down any, I took my gun and was busy clipping them from the branches with ball when eight Indians came at the report of my gun. They were all painted with red earth, armed with bows, arrows, spears of bone, and flint knives, and seemed to me anything but friendly. I endeavoured to explain to them what

I wanted and they seemed satisfied and sat down to smoke, but had no sooner done so than I perceived one string his bow and another sharpen his flint knife with a pair of wooden pincers and hang it on the wrist of the right hand, which gave me ample testimony of their inclination. To save myself I could not do by flight, and without any hesitation I went backwards six paces and cocked my gun, and then pulled from my belt one of my pistols, which I held in my left hand. I was determined to fight for life. As I as much as possible endeavoured to preserve my coolness and perhaps did so, I stood eight or ten minutes looking at them and they at me without a word passing, till one at last, who seemed to be the leader, made a sign for tobacco, which I said they should get on condition of going and fetching me some cones. They went, and as soon as out of sight I picked up my three cones and a few twigs, and made a quick retreat to my camp, which I gained at dusk. The Indian who undertook to be my last guide I sent off, lest he should betray me. Wood of the pine fine, and very heavy; leaves short, in five, with a very short sheath bright green; cones, one 14½ inches long, one 14, and one 13½, and all containing fine seed. A little before this the cones are gathered by the Indians, roasted on the embers, quartered, and the seeds shaken out, which are then dried before the fire and pounded into a sort of flour, and sometimes eaten round (sic). How irksome a night is to such a one as me under my circumstances! Cannot speak a word to my guide, not a book to read, constantly in expectation of an attack, and the position I am now in is lying on the grass with my gun beside me, writing by the light of my Columbian candle — namely, a piece of wood containing rosin.

Friday 27th October My last guide went out at midnight in quest of trout with a flare and brought one small one in the morning, which I roasted for breakfast. He came two hours before day in great terror and hurry and uttered a shriek. I sprang to my feet, thinking the Indians I saw yesterday had found me out, but by gesture I learned he had been attacked by a large grizzly bear. I signed to him to wait for day, and perhaps I would go and kill it. A little before day Bruin had the boldness to pay me a visit, accompanied by two whelps, one of last year's and one of this. As I could not consistently with my safety receive them so early in the morning, I waited daylight and accordingly did so. I had all my articles in the saddle-bags and the horse a mile from the camp, when I mounted

my own, which stands fire admirably and rode back and found
the three feeding on acorns under the shade of a large oak.
I allowed the horse to walk slowly up to within twenty yards,
when they all stood up and growled at me. I levelled my gun
at the heart of the mother, but as she was protecting one of the
young keeping them right before her and one standing before
her belly, my ball entered the palate of the young one and came
out at the back part of the head. It dropped instantly, and as
the mother stood up a second time I lodged a ball in her chest,
which on receiving she abandoned the remaining live young and
fled to an adjoining hummock of wood. The wound was mortal,
as they never leave their young until ready to sink. With the
carcase of the young one I paid my last guide, who seemed
to lay great store by it. I abandoned the chase and thought it
prudent from what happened yesterday to bend my steps back
again without delay. So I returned and crossed the river two
miles further down, and camped for the night in a low point of
wood near a small stream. Heavy rain throughout the day.

Saturday and Sunday 28th and 29th Both days very rainy, and
having very little clothing and impossible to keep myself dry
night or day obliged me to make all the exertion in my power
to reach the camp near the sea, and being under the necessity
of leading my horse the whole distance, he being greatly fatigued
and the road daily getting worse by the continual rain. Camped
on Saturday evening at my second crossing-place, but could get
no food from the Indians, the bad weather preventing them from
fishing, the river being much swollen. Boiled the last of my rice
for supper, without salt or anything else, and had but a scanty
meal. At daybreak on Sunday I resumed my march and went on
prosperously until I came to the large woody hill half-way
between our first march on the north bank of the river, when one
of the poor horses fell and descended the whole height over the
dead wood and large stones, and would have been inevitably
dashed to pieces in the river had he not been arrested by being
wedged fast between two large trees that were lying across the
hill. I immediately tied his legs and head close to the ground to
keep him from wrestling, and with my hatchet I cut the lower
tree and relieved him, having received but little injury. I felt
over this occasion much, for I got him from Mr McLoughlin
and it was his favourite horse. Reached the camp in the dusk,
where I found Mr Michael Laframboise, the Chenook interpreter, and an Indian boy, who told me the Indians had been

troublesome since the brigade of hunters left him on Monday. He kindly assisted me in pitching my tent, gave me a little weak spirit and water, and then made a basin of tea, which I found very refreshing. Very heavy rain during the night.

Monday 30th Last night about ten o'clock several Indians were seen round our camp all armed, and of course instead of sleep we had to make a large fire, leave the camp a little distance, and hide in the grass to watch. An hour and a half before day a party of fifteen passed us, crashing among the grass towards our fire; we immediately fired blank shot and scared them. Returned to the camp and made some tea and ate a little dry salmon for breakfast, and as I had not a single bit of dry clothing and it still raining, I sat in my tent with a small fire before the door the whole day.

Tuesday 31st Heavy showers, with south west wind off the ocean. Cold and raw. Brought wood in the morning for fuel and some branches of pine and bracken for bedding. At noon an Indian who had undertaken to guide two of the hunters to a small lake twenty or thirty miles to the south east of this, returned to our camp and brought on his back one of their coats and had in his possession some of the hunting implements, and looked altogether very suspicious: for the present, as we do not understand their language, we pay no attention; perhaps he has stolen and not murdered them. Kept up our watch as usual; find myself greatly fatigued and very weak. Were not troubled during the night.

Wednesday November 1st Heavy rain until two o'clock. In the afternoon Baptist Mackay returned from the coast, who tells me he hardly ever experienced such bad weather; he had not a dry day. We felt a little relieved to think our small party getting strong, particularly such a one as Mackay, as he will soon procure us fresh food. Evening cloudy.

Thursday 2nd Baptist Mackay went out in the morning and very fortunately killed a fine large doe Long-tailed Deer, which he brought home on his horse at noon. I was glad to stand cook, and ere 4 p.m. I had a large kettle of fine rice soup made, and, just as we were sitting down to eat, thirteen of the hunters came in sight in five canoes and of course were invited to partake. I find this evening pass away agreeably to the eleven preceeding, and although the society at many times uncouth, yet to have a visage of one's own colour is pleasing; each gave an account of the chase in turn. I find myself stand high among them as a marksman and passable as a hunter.

Saturday 4th I had not been at our camp more than an hour last night when I had the satisfaction of being joined by Mr McLeod from his travels to the southward. He informs me that this river (the Umpqua) at its confluence with the ocean is about three-quarters of a mile broad and has a shallow sand-bar and much broken water at the flow of the tide; will not admit ever of any shipping. He journeyed along the sea-beach for twenty-three miles, when he came to a second river, similar in size to this one and also affording the same sort of salmon and salmon-trout. At its mouth are numerous bays: some of them run considerable distances through the country, which is by no means so mountainous as that northwards, and in one of the said bays he pursued his route to the south in a canoe for twenty miles, where he came to a third river, a little smaller than the others, but by the Indian account takes its waters a long distance in the interior. Abounds with the same fishes. Here for the present his expedition stops until he has his party all forward. By his account from the Indians a large stream of water falls into the sea, perhaps about sixty miles still further to the south, where the natives are said to be very numerous: one of his linguists, who has seen the Columbia and the new river, says it is much larger than it. Mr McLeod tells me the country on the coast assumes a very different appearance from that on the Columbia. All the natives like those here had never before seen such people as we are, and viewed him narrowly and with much curiosity; but hospitable and kind in the extreme. Kindled his fire, assisted in making his encampment, glad and pleased beyond measure on receiving a bead, ring, button, in fact the smallest trifle of European manufacture for their services. Have the same clothing and houses as those in this neighbourhood. Mr McLeod tells me that two of his men are going to Fort Vancouver with a despatch on Monday morning, and as the season is far spent and the rainy weather set in, and at the same time doubtful if he will have any more communication before I should start on March 1st for the other side of the continent, I have made up my mind to return, and shall retain a grateful recollection of the kindness and assistance I have uniformly had from this gentleman. (Recollect on your arrival in London to get him a good rifle gun as a present.)

Monday 6th Heavy rain until noon, with a high westerly wind detained all day in consequence of it.

Tuesday 7th The rain last night fell in torrents but moderated at daybreak. As good weather could not be looked for at this late season of the year, I resolved on beginning my march. Started at ten o'clock a.m. with John Kennedy, an Irishman, and Fannaux, a Canadian, and nine horses. Mr McLeod expressed his regret to see me leave with such a small stock of food and that not of the best quality: a few dried salmon-trout, which were purchased of the Indians, and a small quantity of Indian corn and rice mixed together, which was brought from Fort Vancouver. In all, a week's food for two persons; but at this season I hope to find abundance of wildfowl, failing meeting with small deer, so that there is little to be feared as to starving. As the late rains had rendered the high hill impassable for bonded horses, we were under the necessity of carrying our baggage up the river in three small canoes. Camped twelve miles up the river near two Indian lodges and had from the Indians some salmon-trout. Towards dusk it became fine and fair, with clear moonlight, which gave us an opportunity of drying our clothing.

The had an appalling journey back, for it rained almost incessantly, they were often short of food and the horses suffered terribly.

Saturday 11th Last night, after lying down to sleep, we began to dispute about the road, I affirming we were two or three miles off our way, they that we were quite close to our former encampment; all tenacious of our opinions. The fact plainly this: all hungry and no means of cooking a little of our stocks; travelled thirty-three miles, drenched and bleached with rain and sleet, chilled with a piercing north wind; and then to finish the day experienced the cooling, comfortless consolation of lying down wet without supper or fire. On such occasions I am very liable to become fretful. Before sleeping we had agreed to go to a small lake seven miles further on, next day, where we hoped to find wildfowl and give the horses some rest. At day-break I started on foot for the lake, leaving the men to bring up the horses; but being, as I have already observed, off our way the preceding evening, I had only walked about three miles when I perceived myself again off the road. The day being

cloudy and rainy, and having no compass, I thought it prudent
to return to the camp, which I did and found they had started,
but by which course I could not say. I looked about and readily
found our camp of October 7th and then proceeded by the old
route. About midday I was met by Kennedy, who had gone
to the lake by a new way and not finding me there became
alarmed about my safety, and had come in search of me, leaving
Fannaux to take care of the new camp. On reaching the plain
three miles from the camp at 4 p.m. I proposed to go in search
of wildfowl if he would go and assist Fannaux with the encampment; we did not part without my getting strict caution about
going astray a second time. By six o'clock I had three geese and
one duck, and on my way home, when I observed a large flock
a little to the left of my path, I laid down my hunt, gun-slip, and
hat to approach them, and after securing one returned in search
of my articles, but was unsuccessful in finding them, although I
devoted two hours to it. Reluctantly I gave it up and proceeded
to the camp, and as the night was exceedingly dark I would
have had some difficulty in finding it had they not made signals
with their guns to guide me. Close to the camp fired among a
cloud of ducks that were flying over my head and killed one!
I was hailed to the camp with 'Be seated at the fire, Sir', and
then laughed at for losing myself in the morning, my game and
other property in the evening. There is a curious feeling among
voyageurs. One who complains of hunger or indeed of hardship
of any description, things that in any other country would be
termed extreme misery, is hooted and brow-beaten by the whole
party as a pork-eater or a young voyageur, as they term it; and
although in many instances I have observed they will endure
much privation through laziness, and not unfrequently as a
bravado, to have it said of them they did so-and-so, I found
in this instance my men very willing to cook the fowls and still
less averse to eating them. Heavy rain.

On November 15th they came on the conference of the
Sandiam River, south of present day Salem. It was swollen.

Wednesday 15th   Light rain. In the morning I left the camp at
daylight in search of game, leaving the men with the horses and
being scarce of them and at the same time weak, I chose to
walk. On arriving at Sandiam River, which falls in the
Willamette, a stream of considerable magnitude, we found the
village deserted and no canoes. A raft could soon have been

made, but from the rapidity of the current we could not guide it across. Therefore we looked up and down for the most suitable place to swim. The men chose to swim on their horses, I alone. Fannaux in the midst of the stream in spurring on his horse, imprudently gave the bridle a sudden jerk, when rider and horse went hurling down before the current; fortunately he extricated himself from the stirrups, and of course had to adopt my plan of swimming alone. I had articles of my clothing and my bedding drenched, but what gave me most pain was the whole of my collection being in the same state. Proceeded on and found an Indian village only two miles further on, with plenty of canoes. Camped about three o'clock, being fain to give my collection and clothing time to dry, which employed me all the evening. Killed no game; gained about eighteen miles. Tonight, from constant exposure to the wet and cold, my ankles are swollen, painful, and very stiff.

Fortunately no other mishaps occurred and the party reached Fort Vancouver late on the evening of November 19th. Letters awaited him. There were some new faces at the post, including Lt. Aemelius Simpson R.N., George Barnston and Chief Trader James McMillan, all of whom were to make their mark in the history of the North West.

By December 9th he was off again, this time to visit his friend Cockqua near Gray's Harbour. He was made welcome but food was scarce and foul and even Douglas' iron constitution revolted.

"The salmon is very bad, lean in the extreme, killed in the small creeks in September, October and November in the spawning season: when dried resembles rotten dry pine-bark. Having nothing but this to subsist on, I was seized with a most violent diarrhoea which reduced me in four days unable to walk".

He set off back for the Fort in poor order and was forced to surrender his neckerchief and coat buttons to some villainous Indians for a little fresh sturgeon. He arrived worn out at the post on Christmas Day but by the end of the year had fully recovered.

6

1827

The Hudson's Bay Company Express
Fort Vancouver Across The Rockies

The winter of 1826/1827 was terribly hard and some of the horses died. Douglas shot a Californian condor (now a very rare species) but most of the time was spent packing up in readiness for the H.B.C. express across country to York Factory. This route through the Athabasca Pass had been discovered by David Thompson in 1811 and developed by the North West Company.

Monday 1st January Morning dull but fine. The New Year was ushered in by a discharge of the great guns at daybreak. Day spent much to my satisfaction: after breakfast took a ride on horseback and carried my gun; returned at dusk to dinner. The evening, like many I have passed in N.W. America, lay heavy on my hands.

Tuesday 2nd - Wednesday 31st On the 5th, heavy rain and sleet, with a south east wind, succeeded at night by keen frost, 10° above zero. On the 7th 8th and 9th snow, with little intermission: a regular fall of 18 inches to 2 feet over the whole country. The forest presents a most dismal appearance, the immense pines loaded with snow and their wide, spreading branches breaking under their load. This to me irksome, being prevented from going out, the snow too soft for snowshoes. On the 22nd slight thaws during the day, showers of hail and rain, sometimes freezing at night. To pass away the time I copied some notes of the Chenook tribe of Indians.

Thursday February 1st - Wednesday 28th The changeable weather of the last month continued until the 10th of this, when we had a second fall of snow, 15 inches deep, which lay until the 25th, and after that, frequent rains and gusts of wind. Killed a very large vulture, sex unknown. Obtained the following information concerning this curious bird from Etienne Lucien, one of the hunters who has had ample opportunity of observing them. They build their nests in the thickest part of the forest, invariably choosing the most secret and impenetrable situations and build on the pine-tree a nest of dead sticks and grass; have only two young at a time; eggs very large (fully larger than a goose-egg), nearly a perfect circle and of a uniform jet black. The period of incubation is not exactly known; most likely the same as the eagle. They have young in pairs. During the summer are seen in great numbers on the woody part of the Columbia, from the ocean to the mountains of the Snake River, four hundred miles in the interior. In winter they are less abundant: I think they migrate to the south, as great numbers were seen by myself on the Umpqua river, and south of it by Mr McLeod, whom I accompanied. Feeds on all putrid animal matter and are so ravenous that they will eat until they are unable to fly. Are very shy: can rarely get near enough to kill them with buck-shot; readily taken with a steel trap. Their flight is swift but steady, to appearance seldom moving the wings; keep floating along with the points of the wings curved upwards. Of a blackish-brown with a little white under the wing; head of a deep orange colour; beak of a sulphur-yellow; neck, a yellowish-brown varying in tinge like the common turkey-cock. I have never heard them call except when fighting about food when they jump trailing their wings on the ground, crying 'Crup Cra-a,' something like a common crow. The remainder of the month heavy rains.

Saturday 10th - 19th Learning that the 20th of this month was the day fixed for starting on my journey across the continent, commenced packing my collections to go by sea in the first vessel for England.

Tuesday 20th Showery all day. Preparations being made for the annual express across the continent; by five o'clock in the afternoon I left Fort Vancouver in company with Mr Edward Ermatinger for Hudson's Bay, Messrs McLoughlin, McLeod, Annance, and Pambrun for the interior. We were accompanied to the riverside by the few remaining individuals who constituted my little society during the winter, where we wished each other a long farewell — I glad that the time was come when my steps should once more be bent towards England. I cannot forbear expressing my sincere thanks for the assistance, hospitality, and strict attention to my comfort which I uniformly enjoyed during my stay with them — in particular manner to Mr McLoughlin (Chief Factor). Camped at sundown four miles above the establishment.

Friday 23rd Made the portage over the Lower Dalles by three o'clock, and the Upper or Little Dalles by five, and lest we should be annoyed by the Indians on the Falls, four miles higher up, we camped on the gravelly beach of the river. Wood, being scarce, wherewith to boil our kettle, was purchased. Watched all night.

Saturday 24th Started at five and crossed the Falls portage at nine, where we breakfasted. In the interval the boat was gummed and otherwise repaired, being slightly injured. Camped on the north side of the river, seven miles above Day's River, at dusk. Were joined by our friends, who walked all day. The servant, Overy, who had waited behind for the purpose of bringing up the horses, came to us two hours after dark, having in his hand, eight or ten broken arrows which he wrested from an Indian who threatened to put one through him if he did not allow himself to be pillaged. He might have laid him dead on the spot, but prudently chose to allow him to walk away, being rewarded previously with a heavy flogging and deprived of his bow and arrows. Five horses at the camp and the owner agreed to go to Wallawallah. Very high wind from the south west during the night.

Monday 26th At daylight went off on foot over a point of land and met the boat at the lower end of the big island at nine o'clock, when we took breakfast; and having to cross over to the south channel, the north being too shallow, I embarked for the remainder of the day. Put ashore at two at the upper end of the island, where we discovered that Mr McLoughlin's gun had been left at our breakfast-place, and being loth to lose it, having some celebrity attached to it (Sir Alexander McKenzie used it on both his former journeys), Overy, another Canadian, and an Indian were despatched for it; in the meantime we halted for them. Purchased some horseflesh of the Indian, on which we supped. Very high wind during the whole night. Took my turn of watching and cooking by the kettle with Mr McLeod.

Tuesday 27th At daylight Dupond the Canadian returned, and

H

told us that Indians had been at our breakfast place and carried off the gun, and Overy had gone in quest of them and informed us he would soon follow. With difficulty four horses were hired, of a Kyusse Indian, when about eleven o'clock three started overland, leaving a horse for Overy to follow.

Wednesday 28th Started at five and reached the establishment at eleven, where we found our friends, who came overland and arrived there last night. Stayed until three o'clock, during which time I changed what few plants I had gleaned and put my grouse in order. Camped three miles below the Snake River. Evening fine.

Friday 30th Heavy rain last night, which continued until mid-afternoon. Walked all day; nothing worthy of notice occurred. Saw three grouse which escaped, being unable to keep our guns dry. Camped in the bend of the river, fifteen miles below the Priest Rapids. Much fatigued and my feet painful from the gravel and shattered rocks, and having nothing but shoes of deerskin dressed — that is, the hair off and smoked with rotten wood. Had a fine camp: plenty of firewood, which enabled us to dry our clothing.

Monday 2nd - Thursday 5th The river flowing through a more mountainous country, and further to the north, scarcely a vestige of vegetation can be seen, only the gravelly bank and north side of the river, all the ground covered with snow. Walked along the banks of the river picking up any mineral that seemed curious: found some very fine pebbles. Arrived at Okanagan on Thursday, a little before dusk.

Friday 6th Fine, clear, and pleasant. At two o'clock I alone embarked in the boat to go round the big bend, a day and a half's journey, being much fatigued and my feet very painful, blistered, and blood-run, having walked eleven days. My fellow-travellers remained to come over the point on horseback. Parted with Mr A. McDonald, from whom I have had much information, assistance, and hospitality. Camped fourteen miles above the establishment. I intended to have left the pair of grouse here, but not being perfectly dry I was afraid they would fall a prey to insects.

Friday 13 - Tuesday 17th Weather changeable: hail, snow, and rain, wind northerly. The first night of my arrival, I had the great misfortune to get my pair of grouse devoured, the skins torn to pieces by the famished Indian dogs of the place. Although they were closely tied in a small oilcloth and hung from the tent-poles, the dogs gnawed and ate the casing, which were leather thongs. Grieved at this beyond measure. Carried the cock bird 457, and the hen 304 miles on my back, and then unfortunately lost them. Wrote a note to Mr Archibald McDonald at Okanagan to endeavour to procure for me a pair against the sailing of the first vessel for England. Mr Work showed me a pair of Mouton Blanche of the voyageurs, male and female, skins in a good state of preservation.

By mid-afternoon of Tuesday, preparations being made for our departure, I in company with Mr McLoughlin and Mr McLeod took an airing on horseback and returned at dusk to dinner. About nine o'clock at night I was conveyed to my camp, about a mile above the establishment, where we pitched in order that no time would be lost in starting in the morning by them, who spent a few minutes with us and then returned. Having now just bid farewell to my Columbian friends, I cannot in justice to my own feelings refrain from acknowledging the kindness shown to me during my stay among them, a grateful remembrance of which I shall ever cherish. My society now is confined to Mr Edward Ermatinger, a most agreeable young man who goes to Hudson's Bay with us and seven men — four Canadians and three Iroquois Indians. Our next stage is Jasper House in the Rocky Mountains, distant about 370 miles. Laid down to sleep at 2 a.m.

The route from Kettle Falls lay up the Arrow Lakes to the Boat Encampment on the great bend of the Columbia. There their boats were placed en cache and they followed the partly blazed trail to the pass. According to A. G. Harvey the route lay up the Wood River to Jeffrey Creek. The almost freezing river was forded many times and they then had to ascend the ridge of the Big hill. The journey was full of danger and hardship, but tremendously interesting.

On May 1st Ermatinger killed 'a beautiful male partridge' which Douglas successfully skinned and which was later recognised as a new species and called Franklin's grouse after Sir John Franklin. The stuffed bird, beautifully preserved, is now in the Royal Scottish Museum in Edinburgh and I have had the pleasure of handling it. Later that day Douglas climbed on the great peaks — the first man ever to ascend one of the Northern Rockies and although he greatly overestimated the height of the range, it was a most notable first. Indeed no other peaks were attempted for more than thirty years.

Wednesday 18th Overslept ourselves this morning and were not up until daylight, when we hurriedly pushed off lest we should be seen by our old friends, who left us last night. Camped at dusk eight miles above the Dalles on the left-hand side of the river. Travelled twenty-nine miles. Mr Ermatinger, during the time of boiling the kettle, favoured me with some airs on the flute, which he plays with great skill. Noon cloudy.

Thursday 19th Country more mountainous and rugged, the timber smaller. Ten miles from our camp, about eight o'clock a.m. passed Flathead River, a stream not more than 30 yards broad at its entrance, but throws a large body of water into the Columbia. The entrance is cascades, 9 or 10 feet high, over which the water dashed, which has a fine effect, issuing as it were from a subterranean passage, both sides being high hills with large pines overhanging the stream. The headwater of this stream was passed by Lewis and Clarke in their tour across the continent. I am informed by Mr P. Sogden, who possesses more knowledge of the country south of the Columbia than any other person, that its source is a small lake in the Rocky Mountains, which discharges water to both oceans: from the east end is the headwater of one of the branches of the Missouri, and one, as I have observed, is a feeder of the Columbia. Took breakfast two miles and a half above it on the opposite side at nine, where we stayed our usual time, half an hour. From the high grounds on the bank of the river, as far as the eye can behold, nothing is to be seen but huge mountains, ridge towering above ridge in awful grandeur, their summits enwrapped in eternal snow, destitute of timber, and no doubt affording but a scanty verdure of any sort. Lower down the scene is different: rugged perpendicular cliffs of granite and scattered fragments which from time to time have been hurled from their beds in masses too large and weighty for anything to withstand. Travelled thirty miles.

Friday 20th Slight frost in the morning. The tent being wet and partly covered with snow from the preceding night, a small fire had to be kindled in it before it could be folded. Passed, about a mile above our camp, McGillivray's or Cootanie River, also a stream of some magnitude, rapid, and very clear water. This is side to be a good route across the mountains, but from the hostile disposition manifested by the natives inhabiting the higher parts of the Saskatchewan, the Athabasca portage is preferred, being free from such visitors. Five miles above it the Columbia gradually widens to a lake, one to two and a half miles broad, some places very deep, having bold perpendicular rocks; at other places small bays with gravelly or sandy beach with low points of wood. The scenery today is fine, but not so broken, the hills fully as high and more thickly wooded; high snowy peaks are seen in all directions raising their heads to the clouds. Took breakfast at 8 a.m., gained then nine miles. Course of the river then north west and by north west, and north east. About 10 p.m. a light breeze sprang up which enabled us to use a sail, which slackened during the middle of the day, but freshened up again in the afternoon. Camped at dusk on a low sandy point on the left side, four miles from the upper end of the lake. Our distance this day is about forty-seven miles.

Saturday 21st Shortly after dusk last night an Indian and his two children came to our camp and sold a small piece of venison and a few small trout, 10 to 14 inches long, of good quality, and some small suckers, so common in the lower parts of the river. I learn that sturgeon is in the lake, but is not fished by the Indians. Morning clear and fine, wind easterly, which greatly impeded our progress. Started at daylight and continued our route along the north shore. At seven passed a camp of Indians, consisting of three families, from whom three pair of snowshoes, such as I obtained at Kettle Falls, were purchased. Reindeer (Cariboux of the voyageurs) it would appear are found in abundance in the mountains: not fewer than a hundred skins were in this lodge. They are killed readily during the deep snow with the bow. The large hoof which this species has (not observed in any other of the genus) is a proof of the wise economy of Nature, given it to facilitate its tedious wanderings in the deep snows. At 10 a.m. put ashore at the upper end of the lake to breakfast, where we stayed three quarters of an hour. Instead of four miles, as I observed last night, I found it to be eleven. Here were four Indians gathering from the pines a species of lichen, of which they make a sort of breadcake in times of scarcity. In their camp were horns of Blacktailed deer and one pair of Red, or stag, the first I have seen since I left the coast. The canoes of the natives here are different in form from any I have seen before; the under part is made of the fine bark of *Tsuga,* and about 1 foot from the gunwale of birch-bark, sewed with the roots of *Thuya,* and the seams neatly gummed with resin from the pine. They are 10 to 14 feet long terminating at both ends sharply and are bent inwards so much at the mouth that a man of middle size has some difficulty in placing himself in them. One that will carry six persons and their provisions may be carried on the shoulder with little trouble. Weather at noon pleasant; thermometer 55° in the shade; chilly towards evening. Camped on the right near a high rock of pure white marble. A mile on each side of the lake stumps and entire dead trees stand erect out of the water; by some change in Nature the river has widened. The same thing occurs ten miles above the Grand Rapids, 148 miles from the sea. Our distance today is thirty-one miles.

Sunday 22nd In the grey of the morn, at four o'clock, we were on the water and pleasantly pursuing our journey, it being calm and the lake fine and smooth. Slight frost during the night; noon fine and warm but cloudy, which continued throughout the day. Crossed over the right side and passed two points a little beyond the latter in a gravelly bay where we stopped to breakfast at nine, having already gained fifteen miles. This part of the lake has bold rocky shores. Course north and by east. Four miles further on the right there is a remarkable rock 240 feet high, perpendicular, of blue limestone on a substratum of granite. From this point one of the most sublime views presents itself: nine miles of water about five miles in breadth having on the left a projecting point resembling an island and a deep bay on the right, with lofty snowy peaks in all directions.

At the end of the lake, at the foot of a high and steep hill, were three Indian lodges. Camped on the land. Purchased of them a little dried reindeer-meat and a little black bear, of which we have just made a comfortable supper. They seem to live comfortably, many skins of Black-tailed, Rein, and Red deer being in their possession. I purchased a little wool of Mouton Blanche as a specimen of the quality of the wool; gave seven balls and the same number of charges of powder for it. (Get a pair of stockings made of it.) Camped at dusk on a high point of wood (the channel of the river being here covered with snow) on the right hand. A Sunday in any part of Great Britain is spent differently from what I have had in my power to do.

Day after day without any observance (except date) passes, but not one passes without thoughts of home.

Monday 23rd As usual, started in the grey of the morning, about four o'clock. Banks of the river low and shallow with very gravelly banks; on the low points *Thuya* predominates, some enormously tall. I measured one 157 feet and another 204 feet. Breakfasted on the right-hand side at nine; gained nine miles. Purchased some fish of a woman, consisting of three kinds — grey and red suckers, and white mullet, the latter of fine quality. Continued our journey at ten. The river and country the same until five o'clock p.m., being then about fifteen miles further, where the river takes a sudden bend to the north-east and to all appearance loses itself in the mountains. At this place and for two miles higher a scene of the most terrific grandeur presents itself; the river is confined to the breadth of 35 yards — rapids, whirlpools, and still basins, the water of a deep dark hue, except when agitated. On both sides high hills with rugged rocks covered with dead trees, the roots of which being laid bare by the torrents are blown down by the wind, bringing with them blocks of granite attached to their roots in large masses, spreading devastation before them. Passing this place just as the sun was tipping the mountains and his feeble rays now and then seen through the shady forests, imparts a melancholy sensation of no ordinary description, filling the mind with awe on beholding this picture of gloomy wildness. At the head of this (I would almost say subterranean) passage there is a very dangerous rapid, where the water falls 9 feet over large stones, to pass which took all our united strength: two in the boat guiding her with poles and seven on the line. Carried all my articles lest evil should overtake them. Here it becomes a little broader, the shores also rocky, and owing to the deep snow is dangerous and fatiguing to walk. This night from exertion I can hardly write. No new plants today. Progress twenty-eight miles.

Tuesday 24th Saw two beavers in the water, neither of which could be killed. Likewise a few geese and some ducks of two species, both plentiful on the coast.

Wednesday 25th The rain last night ceased about twelve o'clock. Morning dull but pleasant; the mornings and evenings appear long. The high mountains on the banks of the river screen the sun's cheering influence from us until eight a.m. which is withdrawn shortly after four p.m. Stayed for breakfast at nine at the foot of the Dalles des Morts; gained eight and a half miles; general course, west of north. Here the river is confined to the space of 35 to 50 yards broad, the current exceedingly rapid and obstructed by large stones; one place in the middle of the narrows it is dashed over the shattered rocks, which it leaves in foam for the distance of 40 yards below. Reached the head of the Dalles, about a mile and a half long, shortly after noon, but not without much labour and anxiety. Carried all my valuables (seeds and notes) on my back along the rocks; my other articles brought by one of the men. This place, which is looked on as one of the most dangerous parts of the whole river, derives its name from a melancholy circumstance which occurred a few years ago to a party of men who were ascending as we now are. They had the misfortune to have their craft dashed to pieces among the rocks when by a supernatural exertion all escaped to shore, where they endured a short respite from death, more unsupportable than immediate death itself: without food, without arms, scarcely any clothing, being stripped, and three hundred miles from any assistance. In that state, between the hope of life and dread of death, which must be trying to all mankind, two agreed to make an effort to save themselves by endeavouring, before their strength would fail them, to reach Spokane, the nearest establishment, which they did in such state as readily bespoke the misery they had endured. One of the veterans I have seen, an Iroquois, by name Francis, one of the best boatmen. The remaining six being divided in opinion could come to no resolution as to what step should be taken, and no doubt, as is the case in such trying circumstances, became insensible to their safety. All died except one, who, it is supposed on good foundation, supported his dreadful existence on a forbidden fare, having previously imbrued his hands in the blood of his companions or companion in suffering. This, be it as it may, could not be brought home to him in point of law, and the wretch was sent out of their service to Canada. Camped on the right hand side of the river on a sandbank, having gained ten and a half miles above the Dalles. Progress nineteen miles. Nothing in the way of plants this day. The rocks, micaceous granite. Warm during the day, evening cool; high snowy mountains in the distance at forty miles north. Will prove, I hope, the dividing ridge of the continent.

Thursday 26th **Keen** frost last night; obliged to rise twice to make fire. Experienced a most violent headache the greater part of the day, occasioned by the cold during the night. Walked until in a state of perspiration, which gave me relief. Minimum heat 27°.

Friday 27th   Keen frost last night, morning clear, wind easterly. Started at daylight, having enjoyed but a comfortless night's rest among the stones on the shore of the river. Course north; country the same as yesterday. Breakfasted on the left side at nine; gained eight miles. After resuming our journey for 300 yards, at a short turn of the river one of the most magnificent prospects in Nature opened to our view. The daily wished for dividing ridge of the continent, bearing north-east, distant six miles. The sight of the mountains is most impressive. Their height from the level of the river from 6000 to 6500 feet, two thirds covered with wood, gradually diminishing to mere shrubs towards the confines of eternal snow. One rugged beyond all description, rising into sharp rugged peaks; many beyond the power of man to ascend, being perpendicular black rocks distinctly seen, having no snow on their surface. On the right, rising from the bed of Canoe River, the northern branch of the Columbia, they seem to be most rugged; on the left, rising from the bosom of the Columbia, stands a peak much higher than the former, with a smooth surface. Although I have been travelling for the last fifteen days surrounded by high snowy mountains, and the eye has become familiar to them and apt to lose that exalted idea of their magnitude, yet on beholding those mentioned impresses on the mind a feeling beyond what I can express, I would say a feeling of horror. Arrived at the boat encampment 12 a.m. a low point in the angle between the two branches, the Columbia flowing from the east and Canoe from the north; the former sixty yards wide, the latter forty, but very rapid.

Examined the seeds in my tin-box and found them in good order. Repacked them without delay and at the same time tied up all my wardrobe, toilet, &c., which is as follows: four shirts (two linen and two flannel), three handkerchiefs, two pair stockings, a drab cloth jacket, vest and trousers of the same, one pair tartan trousers, vest and coat; bedding, one blanket; seven pairs of deer-skin shoes, or as they are called, moccasins; one razor, soap box, brush, strop, and one towel, with half a cake of Windsor soap. In addition to these I was presented with a pair of leggings by Mr Ermatinger, made out of the sleeves of

an old blanket-coat or capot of the voyageurs. This trifling as
it may appear, I esteem in my present circumstances as very
valuable. When the half of these my sole property is on my
back, the remainder is tied in a handkerchief of the common
sort. Now that I conceive my wanderings on the Columbia and
through the various parts west of the Rocky Mountains to be
over, I shall just state as near as possible their extent:

In 1825

	Miles
From the ocean to Fort Vancouver on my arrival in April	90
In May, to and from the ocean to Fort Vancouver	180
In June, to and from the Great Falls	210
In July, to and from the ocean and along the coast	216
In August, journey on the Willamette River	133
In September, to the Grand Rapids	96
On the mountains of the Grand Rapids	47
In October and November, to the sea	90
In the same, trip to Cheecheeler River or Whitbey Harbour	53
Ascending said river	65
Portage from it to the Cowalidsk River	35
Descending the latter	40
An allowance of my daily wanderings from Fort Vancouver, my headquarters	850
	2105

In 1826

	Miles
In March and April, from Vancouver to the Kettle Falls	620
In May, journey to Spokane	150
In June, from Kettle Falls to the junction of Lewis & Clarke's River	414
In June, journey to the Blue Mountains	190
In July, a second to the same	137
In July, ascending Lewis and Clarke's River to the north and south branch	140
A third journey to the Blue Mountains from that place	103
From Snake River to Spokane	165
From Spokane to Kettle Falls	75
In August, from Kettle Falls to Okanagan by land	130
From Okanagan to Fort Vancouver	490

In September, October, and November, from the Columbia to the Umpqua River and the country contiguous thereto   593

To the ocean and the bays north of the Columbia in December   125

Daily allowance from my places of rendezvous   600

                                                                         ———

                                                                         3932

In 1827

In March and April, the whole chain of the Columbia from the ocean to the Rocky Mountains   995

                                                                         ———

                                                       Total   7032

My notes will show by what means it was gained.

The account that follows, of the traverse of the now disused Athabasca Pass, is of great interest.

Leaving their canoe at Boat Encampment they pressed on up the valley of the Wood River and on May 1st Douglas made the first recorded ascent of one of the Canadian Rockies which he later christened Mt. Brown. He greatly overestimated its altitude and the altitude of the great mountain he called Mount Hooker, probably because he was misinformed about the height of the summit of the pass. They passed the famous little lake called the Committee's Punchbowl, whose waters flow into both the Pacific and the Arctic, and went on down the Whirlpool river past McGillivray's Rock to the old wooden outpost at Jasper.

Saturday 28th Last night cold; minimum heat 18°, maximum, 51°; obliged to rise during the night to make fire. Delayed commencing my journey, Mr E. being employed laying the boat and other articles en cache, until 8 a.m. when we breakfasted and took leave of the main body of the Columbia in a due east course. Passed a low point of wood of a mile and entered a swamp about three miles long, frequently sinking to the knees in water, which was doubly fatiguing from the thin ice on its surface, too weak to bear us up. Crossed a deep muddy creek and entered a second point of wood of an uneven hilly surface. At eleven obliged to have recourse to my bears' paws or snowshoes, the snow 4 to 7 feet deep, being then soft by the sun's influence.

Much annoyed throughout the day by their lacing or knotting
becoming slack by the wet, and being little skilled in the use
of them, now and then I was falling head over heels, sinking
one leg, stumbling with the other; they sometimes turning back-
side foremost when they became entangled in the thick brushwood. Passed several rivulets which were only seen at the rapid
places, all the other places being covered with snow. Of animals
saw a small Bunting, the whole body of a uniform light brown,
except the wings, which were a dirty white; beak short, thick,
white. Also the Blue-crested Jay, so common on the coast. Saw
two Squirrels about the size of the English one, of a light
chocolate colour, feeding on the seeds of the pines. A large
Wolverene visited our camp in the evening, but escaped before
a shot could be put his way. Today is a scene of some curiosity
even to myself, and I can hardly imagine what a stranger would
think to see nine men, each with his load on his back (food and
clothing), his snow-shoes in his hand, starting on a journey over
such an inhospitable country: one falling, a second helping him
up, a third lagging and far behind, a fourth resting smoking his
pipe, and so on. Mr Ermatinger handsomely offered that all
my articles should be carried and I to go light. This I could not
accept seeing him with his load, and although I was perfectly
satisfied as to their safety, yet I could not but carry what has
already cost me some labour and anxiety. I therefore took all
the seeds in the tin-box and journals, secured in an oilcloth,
weighing 43 lb, my wardrobe and blanket carried by one of the
men. Somewhat tired, my shoulders painful from the straps.
Evening fine.

Sunday 29th Morning clear, minimum heat 23°, maximum 43°.
Started at four, being refreshed by a second sleep, in an easterly
course for six miles, in the course of which made seven traverses
across the river from one point to another on the channel,
which is from three to five miles broad, and at high water
during the summer forms an inland sea covering the whole
valley from the foot of the mountains. Turned to the north
east four and a half miles over the same sort of ground, making
seven more fords: water in several 2½ to 3½ feet deep, current
swift and strong, but not rapid. Did not require snowshoes, the
snow being hard with a strong crust of frost. Found the cold
piercing, alternately plunging to the middle in water 35°
Fahrenheit and skipping with my load to recover my heat among
the hoar frost. At 9 a.m. entered a point of wood where the

snow was 4 to 7 feet deep, with a weak crust not strong enough to support us. Obliged to put on my bears paws; path rough, and in addition to the slender crust, which gives the traveller more labour, were dead trees and brushwood lying in all directions, among which I was frequently caught. Towards noon, the snow having become soft and we weary with fatigue, camped on the brink of a river, where no time was lost in making a little breakfast, every person's appetite being well sharpened by our walk. If good weather visits us, we are thankful; if bad, we make the best of a bad situation by creeping each under his blanket and, when wet, dry it at the fire.

Monday 30th In the grey of the morning we resumed our route on snowshoes in the wood about three-quarters of a mile. Entered a second valley, course north-east. Rested after having travelled two and a quarter miles, in the course of which we made seven fordings over the same river that we crossed yesterday. Continued in the same course for the distance of four miles more until reaching the east end, making four fordings more. Here the stream divides into two branches, that on the left flowing from the north, that on the right due east. Took our course in the angle between the two, north-east, entering a thick wood of the same kinds of timber already noticed. *A. amabilis* more abundant and of greater size. After passing a half mile in the wood reached the foot of what is called the Big Hill, also thickly wooded. Steep and very fatiguing to ascend, the snow 4 to 6 feet deep in the higher spots. The ravines or gullies unmeasurable, and towards noon becoming soft, sinking, ascending two steps and sometimes sliding back three, the snow-shoes twisting and throwing the weary traveller down (and I speak as I feel) so feeble that lie I must among the snow, like a broken-down waggon-horse entangled in his harnessing, weltering to rescue myself. Obliged to camp at noon, two miles up the hill, all being weary. No water; melted snow, which makes good tea; find no fault with the food, glad of anything. The remainder of the day is spent as follows: On arriving at a camp, one gathers a few dry twigs and makes fire, two or three procuring fuel for the night, and as many more gathering green soft branches of *Abies* or *Tsuga* to sleep on, termed 'flooring the house', each hanging up his wet clothing to the fire, repairing snow-shoes, and arranging his load for the ensuing day, that no time may be lost; in the morning, rise shake the blanket, tie it on the top, and then try who is to be at the next stage first. Dreamed last night of being in Regent Street, London! Yet far distant. Progress nine miles.

Tuesday May 1st This morning our fire that was kindled on the snow had sunk into a hole 6 feet deep, making a natural kitchen. Minimum heat 2°, maximum 44°, on the highest part of the big hill. Started at daybreak, finding the snow deeper and the trees gradually diminish towards the summit; laborious to ascend. Went frequently off the path in consequence of not seeing the marks on the trees, being covered with the snow. Reached the top at ten, three miles, where we made a short stay to rest. Course north-east. Descended in the same direction and came on the river which we left two days before. Passed in the valley two small spots clear of wood and one low point of wood of small trees, where we camped at midday, being unable to proceed further from the deep soft snow. Progress seven miles. Mr E. killed on the height of land a most beautiful partridge, a curious species; small; neck and breast jet black; back of a lighter hue; belly and under the tail grey, mottled with pure white; beak black; above the eye bright scarlet, which it raises on each side of the head, screening the few feathers on the crown; resembles a small well-crested domesticated fowl; leaves of pines in the crop. This is the sort of bird mentioned to me by Mr McLeod as inhabiting the higher parts of Peace and Smoky Rivers. This, however, is not so large as described. Perhaps there may be two varieties. Said also to be found in Western Caledonia. This being the first I have seen, could not resist the temptation of preserving it, although mutilated in the legs and in any circumstances little chance of being able to carry it, let alone being in a good state. The flesh of the partridge remarkably tender when new killed, like game that has been killed several days; instead of being white, of a darkish cast. After breakfast at one o'clock, being as I conceive on the highest part of the route, I became desirous of ascending one of the peaks, and accordingly I set out alone on snowshoes to that on the left hand or west side, being to all appearance the highest. The labour of ascending the lower part, which is covered with pines is great beyond description, sinking on many occasions to the middle. Half-way up vegetation ceases entirely, not so much as a vestige of moss or lichen on the stones. Here I found it less laborious as I walked on the hard crust. One-third from the summit it becomes a mountain of pure ice, sealed far over by Nature's hand as a

momentous work of Nature's God. The height from its base may
be about 5500 feet; timber, 2750 feet; a few mosses and lichen,
500 more; 1000 feet of perpetual snow; the remainder, towards
the top 1250, as I have said, glacier with a thin covering of
snow on it. The ascent took me five hours; descending only
one and a quarter. Places where the descent was gradual, I
tied my shoes together, making them carry me in turn as a
sledge. Sometimes I came down at one spell 500 to 700 feet in
the space of one minute and a half. I remained twenty minutes,
my thermometer standing at 18°; night closing fast in on me, and
no means of fire, I was reluctantly forced to descend. The
sensation I felt is beyond what I can give utterance to. Nothing,
as far as the eye could perceive, but mountains such as I was
on, and many higher, some rugged beyond any description,
striking the mind with horror blended with a sense of the
wondrous works of the Almighty. The aerial tints of the snow,
the heavenly azure of the solid glaciers, the rainbow-like hues
of their broken fragments, and huge mossy icicles hanging from
the perpendicular rocks with the snow sliding from the steep
southern rocks, with amazing velocity, producing a crash and
grumbling like the shock of an earthquake, the echo of which
resounding in the valley for several minutes.

Wednesday 2nd My ankles and knees pained me so much
from exertion that my sleep was short and interrupted. Rose
at 3 a.m. and had fire kindled; thermometer 20°. Started at a
quarter-past four through a gradually rising point of wood which
terminated three yards below the highest part of the pass in the
valley. An hour's walking took us to one of the head springs
of the Columbia, a small lake or basin twenty yards in diameter,
circular, which divides its waters, half flowing to the Pacific and
half to the hyperborean sea — namely the headwaters of the
Athabasca River. A small lake about 47° of N. latitude, divides
its waters between the Columbia and one of the branches of
the Missamac, which is singular. This being a half-way house
or stage, I willingly quickened my pace, now descending on
the east side. This little river in the course of a few miles
assumes a considerable size and is very varied. There are two
passes, one four miles from its source and one seven, when
it finds its way over cascades, confined falls, and cauldrons of
fine white and blue lime-stone and columns of basalt, like the
feeders of the Columbia at the deep passes of the mountains;
where the torrents descend with furious rapidity it spreads out

into a broad channel bounded by the mountains. The descent
from the east is much greater than from the west, the mountains
more abrupt and equally rugged. Found the snow eight miles
below the ridge gradually diminish. The heat increases and the
quantity of snow on the east not equal to the west. Passed
on the right a very high (perhaps 400 feet) perpendicular rock
with a flat top, and three miles lower down on the same side
two higher ones, rising to peaks about a mile apart at the base
with a high background which appears two-thirds glacier, and
in the valley or bosom of the three, columns and pillars of ice
running out in all the ramifications of the Corinthian order.
From the mouth of the valley of this awesome spectacle a passage is seen more like the crater of a volcano than anything else:
stones of several tons weight are carried across the valley by the
force of the current during the summer months. The change of
climate is great and the herbage equally so. No more huge
conifers, *Acer* nor *Berberis,* so abundant only a few miles on
the other side. Small pines here take the lead, a few bare species
of *Salix* and under the shade *Ledum palustre* with carpets of
Sphagnum. In the dry hilly parts *Ledum buxifolium, Arbutus
Uva-ursi.* The only bird seen on the extreme height of land is a
small light Jay, who, with all the impudence peculiar to most
of his kindred, fluttered round our camp last night picking any
food thrown to him. Of the structure of the mountains I cannot
speak; it is worthy of notice, however, that all I have yet seen
here and west of the Rocky Mountains have a dip of from 30°
to 45° south-west. I do not recollect a single exception. Blue
and micaceous granits, limestone trap, and basalt are the most
common. Sand or freestone I have not yet seen. Halted at 10
a.m. to breakfast, and the snow being here now, fifteen miles
from the ridge, intend to go on in the evening. At 2 p.m.
started; very warm, 57°. Passed through the valley for three
miles and then entered a rocky point of wood; the river confined
into a very narrow space and rapid. On the dry gravelly shores
Dryas octopetala and another species with narrow entire leaves:
perhaps this may be *D. integrifolia.* Low wood, very wet and
difficult to pass over, *Aralia sp.,* a low shrub, 2 feet high. Went
off my way, being before the others three miles, but as fortune
would have it, just as the sun was creeping behind the hill, I
observed smoke about a mile east of me. Without any loss of
time I soon made up to it and found Jacques Cardinal with
eight horses who had come to meet us. An hour after one of
the men came up, and shortly afterwards we heard several shots

fired, which I knew to be signals for me, which obliged me to
send the man on horseback to say I had arrived at the Moose
encampment. Old Cardinal roasted, on a stack before the fire,
a shoulder of Mouton gris, which I found very fine. He had a
pint copper kettle patched in an ingenious manner, in which he
was boiling a little for himself; this with a knife was all the
cooking-utensils. He observed he had no spirit to give me, but
turning round and pointing to the river he said 'This is my barrel
and it is always running'. So having nothing to drink out of, I
had to take my shoulder of mountain sheep and move to the
brook, helping myself as I found it necessary. Informed that Dr
Richardson had in February arrived at Cumberland House; that
Captain Franklin had met a ship in the North Sea; and that Mr
Drummond, who spent last summer in this neighbourhood, had
in November gone to Fort Edmonton on the Saskatchewan River.
Finding none of my travellers come up, Cardinal gave me his
blanket, reserving for himself the skin of a Reindeer. Mountains
on all sides still as high and uneven, but with less snow; no
glaciers and more wood. Crossed the river fifteen times in three
places; two half full of water, very rapid and full of large stones.
This day marched twenty-five miles.

Thursday 3rd Shortly after daylight Cardinal went with his
horses and brought up my companions at seven o'clock, when
we took breakfast and had our little articles tied on two of the
horses, and proceeded on the bed of the river where it was
still covered with ice and snow, over points of wood, low marshes,
and low hills. The path was extremely difficult from the dead
wood and the ground still soft from the melting of the snow.
General course north-east, at our camp to the main branch of
the Athabasca River, a rapid stream seventy yards broad, where
it is joined by the one on the banks of which we had descended.
Crossed it at the junction, the course being then north; descended
the east bank. Intended to put up at the usual camp, but finding
the horses and land better than expectations, they proceeded to
the end of the portage on horses, I following with the gun in
search of birds. Arrived at a small hut called the Rocky Mountain House at half-past six o'clock much fatigued. Progress this
day thirty-four miles.

Friday 4th This morning I was glad and somewhat relieved to
know that the mountain portage was completed and our journey
for three days would be water communications — namely, to

I

(Fort) Assiniboine. Embarked at daybreak in two birch canoes, and being light went down the stream rapidly. The river banks are low, many places narrow, and widens out to long narrow shallow lakes full of sand shoals. Mountains gradually become lower, more even, and more thickly wooded. Took breakfast on a small low sandy island in the upper lake, where we were joined by a hunter having in his possession a very large female sheep, cut up in quarters, only killed about an hour before on the sub-alpine regions of the mountains, where I am informed they are in abundance. Hair short, coarse, and very thick, of a uniform light brown on the neck, head back and sides; belly a dirty-white. I would judge at 170 lb. weight. Continued our route and passed on the right a high rugged range of mountains, and five miles lower on the left some of lower elevation, seemingly the termination of the dividing ridge. Arrived at Jasper House, three small hovels on the left side of the river, at two o'clock, where we put up to refresh ourselves for the remainder of the day. Minimum heat 29°, maximum 61°. Fine and warm. The country to the south undulating and woody; on the north low and hilly, with even surface, also woody, with a most commanding and beautiful view of the Rocky Mountains on the west and east. The difference of climate is great and the total change of verdure impresses on the mind of the traveller an idea of being, as it were, in a different hemisphere more than in a different part of the same continent, and only a hundred miles apart. Obtained from J. Cardinal a pair of ram's horns, which he considers the largest, and the skull, but I regret the lower jaw is wanting. Had some of the much talked of white fish for supper, which I found good, although simply boiled in water, eaten without sauce or seasoning, hunger excepted, not so much as salt, afterwards drinking the liquor in which it was boiled; no bread. Tonight comfortable.

Saturday 5th Last night an old violin was found at our new lodgings, and Mr E's servant being something of a performer nothing less than dancing in the evening would suit them, which they kept up for a few hours. This may serve to show how little they look on hardship when past; only a few days ago, and they were as much depressed as they are now elated. Morning fine; minimum heat 29°, maximum 62°; wind easterly. At daybreak embarked in our canoe and after passing the sandy shores and shoals of the lake went rapidly before the stream. This day admits of little variety. The river is 100 to 140 yards broad, shallow, full of rapids, and although the canoe drew only about nine inches of water, yet ere 2 p.m. we were under the necessity of putting to shore for the purpose of repair. The banks in the upper parts on quitting the mountains are high, gravelly or sandy clay on a stratum of sandstone lower down, as far as we have come. The banks are low and gravelly, covered with only two species of small pines on the shore. *Betula* and *Alnus* of diminutive growth. Last view of the Rocky Mountains closed at 11 a.m. distant forty miles. The water of this river is muddy and small in comparison with the clear majestic Columbia, and it deserves to be noticed that no stream that flows into it has the clear water of itself. This river abounds with geese and ducks: several were killed, and in the evening the Northern Chia, with his wild but mellow voice aroused our camp. The egg of this bird is about the size of a goose egg, greenish-blue. I am in hopes of having some sent from the Columbia. General course of the river south. Camped on the left bank at sundown, having gained ninety-three miles.

7

1827

The Hudson's Bay Company Express Jasper to New York Factory

Although the worst part of the journey was over a good number of excitements lay ahead. The party made good time to Fort Assiniboine but then were forced to wait for provisions before pressing on to Fort Edmonton. Time was not entirely wasted, for Douglas met Chief Factor John Stuart — one of the great figures of the West — who had accompanied Simon Fraser in 1808. Later he met Sir John Franklin and fellow naturalists Thomas Drummond and Dr John Richardson, returning from Franklin's second Arctic expedition. He also encountered the formidable Governor Simpson — possibly the toughest of all the hard men that served the Hudson's Bay Company. Simpson later became one of the most powerful men in the whole of Canada.

Sunday 6th May 1827 Started at sunrise and resumed our voyage with progress until breakfast, having gained about twenty-seven miles. Found the current less and more ice on the banks of the river. Proceeded only three miles further when we overtook Mr George McDougall and four men on their way from Western Caledonia. Had suffered great hardship passing the mountains from hunger; had been nine days coming from Jasper House, which we left yesterday following the ice as it cleared. Obliged to put up with him until four o'clock, when the ice made a rapid move and we embarked and made six miles more and again obliged to put up for the night. Country the same as yesterday, nothing of interest except to myself. Burnt my blanket and great toe at the fire last night.

Monday 7th At daylight started and proceeded a few miles, when we were again detained by the ice. Made a scanty breakfast of some geese killed the preceding day. After remaining two hours and seeing no likelihood of the ice giving way, a short portage was made over the ice into the main channel, which was open and had a very strong current. Proceeded on our voyage until noon, when we were obliged to put to shore, being overtaken by a large flotilla of ice the whole breadth of the river, which continued until two o'clock. Embarked a second time in hopes from the great quantity of ice that passed the river would be cleared below and that we should meet with no other obstruction. At mid-afternoon we passed a large moose-deer standing on the banks of the river; having only one deranged gun in our canoe a sight of him was all that was had — did not seem shy and approached within sixty yards. At sundown arrived at (Fort) Assiniboine, where we were received by Mr Harriott. The whole distance of this river from Jasper House to this place, 184 miles, admits of no variety, seeing one mile gives an idea of the whole: the banks are low marshy or clay mixed with gravel. The wood of diminutive growth — *Pinus alba, Alnus, Betula.* The underwood *Corylus* and *Mespilus, Neottia, Linnaea* and *Cypripelium.*

Tuesday 8th Provisions being scarce and from the hostile disposition of the Indian tribes in the south it was deemed unsafe to go in so small a party, we intended to wait for the people from Lesser Slave Lake who are hourly expected. As Mr McDougall was going down the river with the intention of procuring food for the men from Columbia and Western Caledonia, I accepted an invitation merely to see the country, being yet too early for affording me any plants, and by doing this I may put myself in possession of some birds. Started at nine o'clock, having a scanty but the best breakfast the place afforded, in one birch canoe and eleven men. Country low and marshy. *Pinus* rare, *Betula* and *Populus* abound.

Wednesday 9th Proceeded on our course, still unsuccessful in procuring food by our own exertions. At eight o'clock came to the camp of a Nipissing hunter who had the day before killed a small black bear, which he gave us with some half-dried beaver-meat, on which we made a hearty meal and then resumed our route down the river as far as the junction of Slave Rise River, which flows out of Lesser Slave Lake into the Athabasca River. We had not been here more than two hours when we were joined by the party from Lesser Slave Lake, headed by

John Stuart, Esq., Chief Factor of the district, who received me in the most friendly manner. During the time the canoes were repairing, a dinner of reindeer steaks was prepared when we embarked and ascended the Athabasca again. Country the same as yesterday, low and marshy.

Thursday 10th Mr Stuart, I find to have a more intimate knowledge of the country than any person I have yet seen, and a good idea of plants and other departments in natural history. He was the first individual who crossed the Rocky Mountains and established Western Caledonia in 1805, and the same year reached the Pacific at Fraser's River near Puget Sound, and has since been over a vast extent of country in these parts, first explored by Sir A. McKenzie. He has been also on the Columbia. He informs me, from a letter received from Dr Richardson, dated Fort Resolution on Great Bear Lake, of the return of the whole expedition without having reached Icy Cape. That a connected survey of 13° of longitude to the west of McKenzie's River has been made, but from the hostile disposition of the Esquimaux they found it impenetrable and that on the navigation opening they should all be at Cumberland. From the information he gives me of the opportunities that canoe travelling affords of collecting subjects of natural history, I have abandoned my idea of going either to Montreal or New York, and agreeably to the plans pointed out by Mr Sabine last year as the better way, I shall sail from Hudson's Bay. By doing this I can remain six weeks at some place in the interior, and still be in time for the ship. By going through Canada nothing could be done and my trip would no doubt be expensive. Perhaps I may be enabled to go to Swan and Red Rivers overland from Carlton House, a journey of twenty days. This I am told will depend on the route taken by the Stone Indians, who are hostile and if in that direction would be deemed unsafe. Learned that Mr McDonald, the person who had in charge my box of seeds addressed to be left at Fort Edmonton on the Saskatchewan River, had endured much misery descending the Athabasca, the ice being taken before he had made good half his journey. In company with him Mr Drummond. Hope my box is safe (do not relish botanist coming in contact with another's gleanings).

Sunday 13th By making an early start ten miles was gained to breakfast; shortly afterwards we left the canoe and cut over a low point of wood and arrived at Assiniboine at two o'clock. Mr

Stuart killed a male partridge, the same species that I saw in the bosom of the mountains, called by him White Flesher — different from the common ruffled grouse — too much destroyed for preserving. Make some small slug and procure a pair of this fine bird.

Tuesday 15th This uninteresting wretched country affording me no plants, at daylight I took a gun and went in quest of partridges; killed a pair of White Fleshers and a hen of a different species. The male a beautiful bird. Only one of the three, the hen of the White Flesher, was worth skinning. Had eggs — small, pure white, about the size of the common pigeon. Pine leaves, and leaves of birch in the crop. The flesh is remarkably white. The breast-bone uncommonly high and very short, differs only from the male in the ruffle being less conspicuous as to colour and size. The hen of the other species was about the same size, of a dark red. Only one shot struck it, but, as misfortune would have it, through the head. Camped at 'Two Rivers' much fatigued.

Thursday 17th Two hours' walk through an excessively bad road took us to Paddle River, a rapid muddy creek, thirty yards wide and at present swollen over its banks from the melting of the snow. A raft was constructed and two men swam across and pulled it over by a rope; by this tedious operation we got all over in the space of three hours. This place affording no fodder for the horses, went on until we came to a low plain at midday, when we, as soon as refreshed in the afternoon, continued through a thick woody country intersected by narrow lakes until we reached Pembina River at dusk, and regretted at not finding the canoes which we parted from six days ago.

Friday 18th Morning fine and pleasant. Three large rafts being made, sufficient to carry all the baggage and persons in one trip, the river being too broad and rapid to return with one, we all crossed except Mr Stuart who intends to await his canoes. Went before the brigade in quest of birds, but still unfortunate. Path a little better than the preceding days; ground high and not so thickly wooded. Shortly after two o'clock arrived at Eagle Lake, six or eight miles long, three to four broad, in which are caught grey sucker, white fish, and pike or jack fish, where we must stay to procure wherewith to complete the remainder of our journey to Saskatchewan. Thirty-one were caught before dusk on the first visit of the net — of pike and sucker but no white fish. Some were also had from a small barrier in the stream flowing out of the lake.

Sunday 20th High wind during the night; minimum heat 36°, maximum 61°. Cloudy towards even. Received a friendly note from Mr Stuart, whom I have just left, that should I arrive at Fort Edmonton and find that a few days can be spent to advantage, to wait and descend the Saskatchewan in his boat. A sufficiency of fish being caught for a part of the journey, raised camp at ten, keeping a south course. Road very bad, much worse than any yet gone over; passing the numerous swamps often sinking nearly to the middle in mud and water. Some of the horses obliged us to camp earlier than expectation, being broken down with fatigue and having to pass a thick wood five miles long where no fodder could be had. Camped on the outskirts on north side. The distance to the establishment being only about forty miles, I intend, should the day be fine to start on foot in order that a few days may be had to collect before the brigade comes up; at the same time I am most anxious to learn the fate of my packet of seeds. Perhaps I may go in one day. This day's march is nine miles.

Monday 21st. At daylight, four a.m., started on foot accompanied by an old Nipissing Indian who had spent many years west of the Rocky Mountains. Although to appearance upwards of seventy years of age, I found him a most excellent walker. Passed a deep muddy swamp a mile broad, and entered a thick point of pine of five more, when I was informed the laborious part of the journey was over. At seven met five men and twenty-five horses going to meet the brigade. They offered me a horse, but I chose to walk, thinking that the horses might be all required to bring up the baggage. Continued my route along Lake Bowland to the south end, where two men had been sent on to fish. Having been unsuccessful and no breakfast, my stay was short. Crossed a deep narrow creek and walked along a low moist meadow through which the just-mentioned creek descended for four miles when the country became very different; a fine undulating ground with clumps of poplar and willow on low parts, *Mespilus canadensis* on the dry spots intermixed with Rose and *Rubus,* both shy in growth, the country being from time to time burned by the Indians. Passed the small deep rivulets by means of throwing down two trees. All the hollow parts of the plains overflowed with water — to all appearance shallow lakes. Appears to have at one time abounded in Red and Long-tailed deer, many horns being strewed over the ground the horns and skulls of buffalo lying in all directions. At three o'clock came to Sturgeon River, a small deep muddy stream but at this season large, the banks overflowed. My hatchet being small, two hours were spent making a raft. I would not have lost three minutes in crossing, but my poor old guide was afraid the chilliness of the water would injure him, having perspired much, and on his account I assisted him in raft-making. Being then only nine miles from Fort Edmonton on the Saskatchewan, my spirits revived and I hastily tripped over the ground and passed many muddy creeks and shallow sheets of water, wading to the middle. Night creeping in on me, my view of the country gradually disappeared. At night, on reaching a rising eminence unexpectedly, I heard the evening howl of the sledge-dogs, which to me was sweet music, and perceived fires in some lodges which I knew to be near the establishment. Being all over with mud, I returned half a mile to a small lake, stripped and plunged myself in and then comforted myself with a clean shirt which I carried on my back in a bundle. I was most kindly received by Mr John Rowand, and had supper prepared for me of fine moose-deer steaks, which were most acceptable after a walk of forty-three miles through a most wretched country without having anything to eat. I found Mr F. McDonald here, who took charge of my box last year. I now learn it had sustained injury, it having been broken. Will see it in the morning. I must mention the particular attention of Mr Rowand, and the manner of conferring it. Thinking that they would be better to have the paper changed, he had them in his presence examined by Mr Drummond, who was there at that time, but who is now at Carlton House. This was kind. Morning frosty; minimum 30°, maximum 49°; showery, with high wind.

Tuesday 22nd   Fatigued so much with yesterday's walking that no sleep could be had; rose at daybreak and had my box opened; found the seeds in much better order than could be expected from the trouble the person had before he reached this place. Only eighteen papers had suffered, amongst which I am exceedingly sorry to say is *Paeonia*. This one of the finest plants in collection. It often happens that the best goes first.

Seeds or plants should be enclosed in soldered tin-boxes to prevent wet or moisture and placed in strong wooden boxes.

Fortunately my shirts were in the box, so they absorbed the moisture. However, from my very small stock being entirely rotten I can at the moment ill spare them.

Wednesday 23rd - Saturday 26th   The country here is undulating, low stunted pines, on the banks of the river, *Populus, Betula* and *Salix,* and a scanty herbage of herbacious plants. A fine young Calumet Eagle, two years old, sex unknown, I had off Mr Rowand; brought from the Cootanie lands situated in the bosom of the Rocky Mountains near the headwaters of the Saskatchewan River. His plumage is much destroyed by the boys, who had deprived him of those in the tail that were just coming to their true colour. Many strange stories are told of this bird as to strength and ferocity, such as carrying off young deer entire, killing full-grown Long-tailed deer, and so on. Certain it is, he is both powerful and ferocious. I have seen all other birds leave their prey on his approach, manifesting the utmost terror. By most of the tribes the tail feathers are highly prized for adorning their war-caps and other garments. The pipe-stem is also decorated with them, hence comes the name. Abundant at all seasons in the Rocky Mountains, and in winter a few are seen on the mountainous country south of the Columbia on the coast. Are caught as follows: a deep pit is dug in the ground, covered over with small sticks, straw, grass, and a thin covering of earth, in which the hunter takes his seat; a large piece of flesh is placed above, having a string tied to it, the other end held in the hand of the person below. The bird on eyeing the prey instantly descends, and while his talons are fastened in the flesh the hunter pulls bird and flesh into the pit. Scarcely an instance is known of failing in the hunt. Its ferocity is equal to the grisly bear's; will die before he lose his prey. The hunter covers his hands and arms with sleeves of strong deerskin leather for the purpose of preventing him from being injured by his claws. They build in the most inaccessible clefts of the rocks; have two young at a time, being found in June and July. This one had been taken only a few days after hatching and is now docile. The boys who have been in the habit of teasing him for some time past have ruffled his temper, I took and caged him with some difficulty. Had a fresh box made for seeds and another for my journals, portfolio, and sundry articles. Could find no lock to put on it. The river here is broad, four hundred yards, high, clayey, and muddy banks water muddy. Coal is found in abundance.

Saturday 26th - Thursday 31st Last night, before we should part with our new friends, Mr Ermatinger was called on to indulge us with a tune on the violin, to which he readily complied. No time was lost in forming a dance; and as I was given to understand it was principally on my account, I could not do less than endeavour to please by jumping, for dance I could not. The evening passed away pleasantly enough; breakfasted at five o'clock and embarked in Messrs Stuart and Rowand's boat with all my baggage and went rapidly before the stream. Day warm and pleasant. Put ashore in the dusk to cook supper, and as the Stone Indians had for the last twelve months manifested hostile intentions it was deemed unsafe to sleep at a camp where fire was. We therefore embarked, and the boats tied two and two together, and drifted all night. Finding this mode of travelling very irksome, never on shore except a short time when cooking breakfast, always dusk before a second meal, I began to think this sort of travelling ill adapted for botanising. Breakfasted at Dogrose Creek, where I found *Ribes hudsonianum* (Richardson in Frankl. 2nd Journ., App. p. 6). The country here changed much for the better; small hills and clumps of poplar and small rocks. Just in the dusk of the evening had a fine chase after two Red Deer swimming in the water, and on following in the boat both were killed; smaller than those west of the mountains. Saw a huge grisly bear (unsuccessful in killing him) and a number of small plain wolves. Passed Fort Vermilion, an abandoned establishment, and Bear and Red Deer Hills, where the country becomes pastoral and highly adorned by nature. Soil dry and light, but not unfertile.

On Wednesday at sunrise five large buffalo bulls were seen standing on a sandbank of the river. Mr Harriott, who is a skilful hunter, debarked and killed two, and wounded two more; all would have fallen had not some of the others imprudently given them the wind, that is on the wind side. Fifty miles further down a herd was seen, and plans laid for hunting in the morning. Some deer were killed this evening and some of the Pronghorned antelope of the plains. Skinned one but unable to preserve the hair on. This little animal is remarkably curious in his disposition; on seeing you he will at first give three or four jumps from you, return slowly up to within a hundred and fifty yards, stand, give a snort, and again jump backwards. A red handkerchief or white shirt — in fact, any vivid colour, will attract them out, and hunters crawl to them on all fours, raising the back like a quadruped walking, and readily kill them.

June 1st   A party of hunters went out at daylight after the herd of animals seen last night. Most willingly I followed them, not for the purpose of hunting but gathering plants. Found *Phlox hoodii* in flower, or rather I might say declining. Laid in specimens — some *Diadelphia* and *Gramineae*. Returned well pleased; supped earlier than usual, and again embarked. Mr Harriot and Ermatinger and three hunters went off to the opposite side to a herd and killed two very large and fine animals. Seeing their boat at the side of the river and no one in it, gave us to know they had all gone for the meat and we put to shore. A party from our boats was sent off to help them. Accompanied by Mr F. McDonald, they readily were guided to their companions by calls, and found Mr H. and B. pursuing a bull that had been wounded, in which he joined. The animal, which had suffered less injury than was expected, turned and gave chase to Mr McDonald and overtook him. His case being dreadful coming in contact with such a formidable animal and exasperated, seeing that it was utterly impossible to escape, he had presence of mind to throw himself on his belly flat on the ground, but this did not save him. He received the first stroke on the back of the right thigh, and pitched in the air several yards. The wound sustained was a dreadful laceration literally laying open the whole back part of the thigh to the bone; received five more blows, at each of which he went senseless. Perceiving the beast preparing to strike him a seventh, he laid hold of his wig (his own words) and hung on; man and bull sank the same instant. His companions had the melancholy sensation of standing to witness their companion mangled and could give no assistance — all their ball being fired. Being under cloud of night, and from what had taken place, his life could not be expected. One returned and acquainted the camp, when each with his gun went off to the spot. On arriving some of the half-breed hunters were in a body to discharge their guns at him, when I called out to Mr Harriot not to allow them to fire all together; that one well-directed shot was enough and by firing more Mr McDonald if alive might fall by one, being close if not under the bull. He agreed to it, but while giving orders to some that he depended on, a shot went off by accident without doing any injury to anyone, and had the unexpected good fortune to raise the bull, first sniffing his victim, turning him gently over, and walking off. I went up to him and found life still apparent, but quite senseless. He had sustained most injury from a blow on the left side, and had it not been for a strong double sealskin shot-pouch, with ball, shot, wadding, &c., which shielded the stroke, unquestionably he must by that alone have been deprived of life, being opposite the heart. The horn went through the pouch, coat, vest, flannel, and cotton shirts, and bruised the skin and broke two ribs. He was bruised all over, but no part materially cut except the thigh — left wrist dislocated. My lancet being always in my pocket like a watch, I had him bled and his wounds bound, when he was carried to the boat; gave twenty-five drops of laudanum and procured sleep. In hopes of finding Dr Richardson no time was lost to convey him to Carlton. The following day several more were killed, but from what I had seen my desire of seeing such dreadful brutes cooled. I continued gleaning plants when the limited times occurred. At 2 p.m. on the 3rd arrived at Carlton House and was received with politeness by Mr Pruden. Here I found Mr Thos. Drummond had come down to meet Dr Richardson in spring. The doctor is now below at Cumberland House. In the evening had an account of his travels and progress and informed me had received a note from Mr Sabine concerning *Phlox hoodii*. He appears to have done well. I must state he liberally showed me a few of the plants in his possession — birds, animals &c., in the most unreserved manner.

Monday June 4th Accompanied by Mr Drummond, I made a trip contiguous to the establishment and was guided to several habitats by him. I learned with regret that my anticipated journey overland to Swan and Red Rivers could not be accomplished. In the first place, two horses would be requisite to carry my papers, blanket, and food — unsafe to have one in the event of dying; in the next place, it was uncertain in what direction the Stone Indians were, and in the event of their meeting me mine would beyond any doubt be a done career. One of the Canadian servants was four weeks ago murdered within four miles of the house, his gun and horses taken, and his body left stripped. The villain who committed this horrid deed was, I am informed, kept during the winter in food, being an object of pity and his family starving; and on his quitting in spring manifested his ingratitude by perpetrating the foulest of crimes. Therefore with regret I had, on the advice of the persons in authority, abandoned it and proceeded to Norway House, where perhaps an opportunity may offer of visiting Red River in preference to going by Canada or the States, both being monstrously costly.

Tuesday 5th - Saturday 9th   The route from Carlton to Cumberland is so well known from the description of the Arctic voyagers that anything from my pen is unnecessary. The journey admits of little variety. Eighty miles below Carlton at a high bank on the left-hand side of the river, called 'The Women's Encampment', the pastoral and rich verdure is no longer seen; instead low thick marshy woods of *Pinus banksiana,* Spruces, Birch, Poplar and Willow — *Carices* and *Gramineae* in the marshes. Very unsteady rainy weather with high winds. During the short turns on the shore picked up a few plants. Arrived at Cumberland at 5 p.m. on Saturday and was kindly welcomed by Mr J. Leith. Here I was greeted by Dr Richardson safe from his second hazardous journey from the shores of the Polar Sea. Every man must feel for the hardship and difficulties which he endured and overcame, and the successful termination of the perilous undertakings. What must be most gratifying is extricating themselves from the formidable Esquimaux without coming to violence. The doctor has a splendid herbarium and superior collection in almost every department of natural history. On telling him I wished to remain at Carlton, he observed that Captain Back would have willingly given me a passage. Informed him of my intention of going to Red River and sailing from Hudson's Bay; approved of it much. Could do nothing going to Canada.

Monday 11th - Saturday 16th   The country throughout presents the same uniformity. Thick low wet woods and muddy banks. Gathered a species of *Salix* and *Carex.* No place for botanising. Towards dusk on the second day reached the head of the Grand Rapid and walked down through the wood while the boats descended — one unfortunately struck on some rocks so that they reached the shore with some difficulty. All next day spent drying the cargo and repairing the boat. Had a ramble in the woods and procured a few things which will be noticed below. Killed a fine large male pelican and preserved the skin. The mischievous boys tore the neck and otherwise injured it. Killed a small plover and preserved it also. High wind and sleet during the whole night and following day. Did not rise until midday. Moderated at sundown, when we embarked and entered Lake Winnipeg. Slept none. Charmed by the mournful cries of the Northern Diver. At eight o'clock put to shore and breakfasted; shore cold, pure white limestone. Found on some of the small rocky islands abundance of gulls' eggs. Towards noon the wind rose very high, producing a high swell, and being near the shore in such broken water, we were under the necessity of lying to. Had much difficulty to land, the water breaking over the boats. Gained the old establishment of Norway House at 1 p.m. where we took some breakfast and at two resumed our route to the new one, eighteen miles below on Jackfish River, where we arrived at 8 p.m. Here I found my old friend Mr John McLeod, who carried my letters across from the Columbia, also J. G. McTavish, Esq., from whom I had much kindness and who desired to know in what way he would be of most service to me. Received a letter of Jos. Sabine, Esq., London, 10th March: good news, the vessel from the Columbia arrived safe; collection sustained no injury. A letter from William Booth, Mr Murray, Dr Hooker of Glasgow, and my brother, the latter affording me but news of a melancholy cast.

June 17th This morning at daylight George Simpson, Esq., (Governor), arrived from Montreal, who I state with pleasure gave sufficient testimony of his friendly attentions and kind offices. Seeing me perhaps rather indifferently clothed, he offered me some linen, &c., which I refused, at the same time extremely indebted to him.

Douglas stayed at Norway House from 16th June to July 2nd overhauling his collection. He was then offered a passage down to the Red River by Sir John Franklin who was moving on to Montreal. They reached Fort Alexander on July 8th and Douglas managed to get a guide down to Fort Garry where he was royally entertained by Governor Donald McKenzie. The great city of Winnipeg has sprung from this nucleus and Douglas thoroughly enjoyed meeting the local settlers and visited schools and churches of all denominations down the Red River. He botanised fruitfully and on August 10th set off northwards again to Norway House, reaching there on August 17th. On August 18th he set off again on the final leg of his journey to York Factory by canoe and portage. The trip was, as ever, not without anxiety and incident but on 28th August he finally reached York Factory: "At sunrise on Tuesday I had the pleasing scene of beholding York Factory two miles distant, with sun glittering on the roofs of the house (being covered with tin) and in the bay riding at anchor, the company's ship from England".

Even then his adventures were not over, for although it is not recorded in his Journal, one last adventure awaited him.

Finding time / hanging on their hands before the ship sailed Douglas, Back and Kendall were rowed out to visit it. On the way back a tremendous storm blew up and they were driven 70 miles out into the Bay, out of sight of land. Their little boat was nearly swamped and they were most fortunate in getting back. Douglas and Kendall did not fully recover their health until they got back to England.

On September 15th the *Prince of Wales* set sail and made an excellent passage arriving at Portsmouth on October 11th.

Monday July 2nd By three o'clock p.m. everything was in readiness for our departure and the canoe in the water, but being lumbered much more than expectation a place for Augustus could not be found; he had therefore to remain at Norway House. Took under my charge a packet of letters for the Red River Settlement and a box containing Church ornaments for the Roman Catholic Bishop. Sent the Calumet Eagle to Hudson's Bay by a Mr Ross, wild fowl and other meat being scarce, and as he will not eat fish I was unable to keep him at the latter place. Placed the white-headed one under the care of a woman attached to the establishment, until my return. Left my sundry articles gleaned in my descent of the Saskatchewan River; and the roots or bulbs brought from the Columbia being still fresh and nearly dry I halved, placing the one in a well-secreted place in the wood, contained in a folded piece of birch bark, fearing the mice may find them; the other in a paper bag, hung up to the roof of the house with some bird-skins. Started at four; pleasant. Camped at 8 p.m. eighteen miles above the establishment. Rain and loud peals of thunder at dusk and during the night.

Sunday 8th Started at 6 a.m. and passed several high limestone cliffs. Took breakfast on a low sandy shore where, in small still waters, was abundance of *Utricularia* in blossom. Remained on shore an hour and then proceeded for twelve or fourteen miles when a stay was made for changing linen. Arrived at the establishment on the River Winnipeg (Fort Alexander, or Basch), the Riviera of the voyageurs. We were welcomed by Mr John McDonald, a brother of the person who crossed the Rocky Mountains last autumn; he was also here on his way to Canada. Became acquainted with the Rev. Mr Picard, of the Roman Catholic Mission at Red River, on his way to Canada.

Monday 9th Early in the morning had a large fire made for drying paper and had all my plants changed before breakfast. Wrote a short letter to Governor Clinton of New York, saying I should sail from Hudson's Bay for England. About 10 a.m. Captain Franklin and Dr Richardson started in their canoe for Canada, and took with them as passenger Mr Picard, he having been disappointed in going by the Company's barges or canoes. Feel obliged to Captain Franklin (good man); will see Mr Sabine. Made several walks round the establishment and collected a few plants. Several heavy showers accompanied by thunder and lightening. The scenery of this place is fine, rich, and very beautiful; well-wooded; low, level country; soil fertile, deep alluvial loam, with a heavy sward of herbage. Requested of Mr John McDonald the favour of hiring me a small canoe to carry me to Red River. Neil McDonald, a person who accompanied Captain Franklin, offered to carry me in his, but found it too small for our luggage. The Indians being camped a considerable distance from the place and all at this season being much engaged, I had hired for me a Canadian, who agreed to carry me for the sum of four dollars and his food. Saw that his canoe was in repair in the course of the evening and made preparations for starting in the morning.

Tuesday 10th High winds during the night and morning from the lake; delayed until ten o'clock, the swell being too heavy for such a small canoe. Being provided with provision for myself and man, I took my leave and descended the river to the lake; pleasant clear weather. Proceeded along the western shore close to the land, which is low, well-wooded; many places overflowed. The lake at this season being high, nothing worthy of notice occurred; saw no plants; observed flocks of passenger pigeons. Camped at dusk on a gravelly beach near a small creek, and was visited by some Indians, of whom I purchased some birch bark for my specimens, serving instead of pasteboard.

Wednesday 11th At six in the morning, I embarked, the wind which blew violently during the night having moderated. About 6 p.m. entered Red River, five or seven miles above the entrance of the lake, having as I stated before cut the overflowed points. The stream is considerable, 250 to 300 yards broad, deep and muddy; banks, low; deep-black alluvial earth; thinly wooded. A few trees of alder, plane, oak and ash along the banks. Laid in some fine strong *Gramineae, Utricularia, Poly-*

K

gonum, and *Asclepias.* Evening close and cloudy. Much annoyed during the forepart of the night by mosquitoes.

Thursday 12th Morning cool, with a heavy dew; started at 3 a.m. up the river. At sunrise passed several thinly planted low houses, with small herds of cattle wandering from the folds; humble and peasant-like as these may appear to many, to me — who have been no sharer of civilised society for a considerable time past — they impart a pleasant sensation. At seven took breakfast two miles below the rapid, where I left the canoe and my luggage to go by land, taking with me my boards and paper. Strangers in this quarter appear to be few; scarcely a house I passed without an invitation to enter, more particularly from the Scottish settlers, who no doubt judging from my coat (being clothed in the Stewart or royal tartan) imagined me a son from the bleak dreary mountains of Scotland, and I had many questions put to me regarding the country, which now they only see through ideal recollections. Appear to live comfortable and have the means of subsistence by little exertion. Walked along the right or north bank of the river; about two o'clock passed the church Missionary establishment and heard the bell ring for the boys to assemble to school; found two at play, one about four years, the other six years of age. I inquired if they were scholars and had answer, 'Yes, sir'. 'What book do you read at school?' The elder, who was the spokesman, said 'I read the parables, and he (pointing to the younger) reads Tom Bowles.' This all pleasing, I presented them with a few trifling articles, when as soon as they manifested their thanks by a low bow they galloped off to their companions, who flocked round them to hear their story. About a mile further on passed a large windmill from which Fort Garry appeared, situated at the junction of Assiniboine River with the river among some wide-spreading oaks, and on the opposite side the Roman Catholic church and Mission establishment, both forming a fine effect. Called at Fort Garry and presented myself to Donald McKenzie, Esq., the Governor of the colony, who received me with great kindness. While a basin of tea was preparing at my request, a large tureen of fine milk was placed on the table, which I found excellent. I handed him Governor Simpson's note, but found that a note was unnecessary with Mr McKenzie. His conversation to me is the more acceptable from the intimate knowledge he possesses of the country west of the Rocky Mountains. In 1819 he ascended the Missouri River and crossed the continent to the mouth of the Columbia with an American party; was the companion of Messrs Nuttall and Bradbury as far as they accompanied the expedition up the former river. He has travelled largely through the country south of the Columbia, in the interior, behind the Spanish Settlements, and like all who share in such undertakings, shared in the fatigues and hardships attendant on these expeditions. But his was more than usual, being the first who ventured on these untrodden wilds. He has since recrossed by the Columbian route. Had a visit paid me by Spokane Garry, an Indian boy, native of the Columbia, who is receiving his education at the Missionary school. He came to inquire of his father and brothers, whom I saw; he speaks good English; his mother tongue (Spokane) he has nearly forgotten. Sent the box and letters under my care to the Bishop. Evening fine.

Saturday 14th Made a short excursion up the Assiniboine River and returned to breakfast. At eleven o'clock Monseigneur J. N. Provenchier, the Bishop of the Roman Catholic Mission, the Rev. Theophilus Harper, and Mr Buchier, a young ecclesiastic, called on me and made a long stay. The Bishop speaks good English, but with that broken accent peculiar to foreigners. Mr Harper speaks the English language with as much fluency as his native tongue, French. They conversed in the most unreserved affable manner and made many inquiries concerning the different countries I had visited. I have some reason to think well of their visit, being the first ever paid to any individual except the officers of the Hudson's Bay Company. I am much delighted with the meek, dignified appearance of the Bishop, a man considerably above six feet and proportionally stout; appears to be a man of the most profound acquirements, seen only through the thick rut of his great modesty. Resumed my walk in the evening, still adding some things to the Collection.

Sunday 15th At church. Heard a sermon from the Rev. David T. Jones, minister of the English church. There being no timepiece for the colony and the habitations widespread, the hour of the day is guessed by the sun. The service being begun half an hour before I got forward, in consequence of missing the proper path, the clergyman, seeing me from one of the windows, despatched a boy to fetch me on the proper path. This struck me as the man of the world who, in the parable, was compelled to go to the feast by the person stationed on the wayside.

After service Mr Jones received me with every demonstration of kindness, and politely invited me to his house and said that I should be no stranger during my stay. Returned home at midday and put some plants in order. Very warm and cloudy.

From the Bishop I had much civility and spent the evening in the most agreeable manner. He showed me his garden, farm, church, and Mission establishmet, which reflects great credit on its conductor. The aborigines and Bulès have not only a religious education but are taught domestic economy, farming, spinning, and weaving cloth from the wool of the buffalo. This establishment, in common with most others, sustained great injury from the high water during last year, and from the pressing state of the church funds the most rigid economy is required to keep the Mission alive.

Monday and Tuesday 23rd and 24th   In company with the Rev. Mr Harper, of the Roman Catholic Mission, who kindly offered to accompany me, I started for the Whitehorse plain, eighteen miles up the Assiniboine River. The country differs in no wise from the country on the banks of the Red River; put up for the night in the house of Mr Grant and were very civilly entertained. Returned at sundown on Tuesday, having made some additions.

Friday 10th   Morning fine. Mr McKenzie having informed me of his intention of sending a boat to Norway House, and lest the others which are to be despatched in a day or two should be delayed in transit by bad weather on the lake too long to meet the ship in Hudson's Bay, I though it prudent to make my stay no longer. The Rev. Mr Jones and the Rev. Mr Cockran called on me and handed me some small packages for Hudson's Bay and for England. Both these gentlemen have shown me much civility. (A few seeds from the Society would be of great benefit to the Missionary establishment and would be thankfully received.) To D. McKenzie (the Governor of the Colony) I am greatly indebted for his polite attentions. After bidding him and the Bishop adieu, I left the establishment in company with Mr Hamlyn, the surgeon, for Hudson's Bay. In our descent of the river we had to make short delays, receiving letters, &c. for Hudson's Bay. Had some cheeses presented me, which I could not well refuse. Called on James Bird, Esq., who resides a considerable way down the river, and received a letter from him addressed to Mr Sabine. I forgot to mention a week ago, being near his house on one of my walks, I called on him and he received me with civility and attention; had tea with him and invited to renew my visit, which I should have done if time permitted. Informed me he had a letter from Mr Sabine last March, mentioning the probability of my calling on him. Camped a few miles below the rapid. Laid in two species of *Artemisia* and a few other plants.

Saturday 18th Left Norway House at six a.m. in company with Mr Jos. Bird, with whom I intend to complete the remainder of my journey, as the other boat was to return to Red River. Passed at eight o'clock two canoes in Play Green Lake containing the men belonging to the Land Arctic Expedition on their way to Montreal. Made but little progress, having a strong wind against us. At midday gained the lower establishment on Jack River, where I found my old friend Mr John McLeod. Learned with regret my Silver-Headed Eagle had died of starvation. I found every other thing safe. The roots, both dry and those hid in the wood, in good condition.

Thursday 23rd Thunder and lightening during the night. As usual made an early start and reached Oxford House at ten, where we took breakfast. Received a letter from Mr Colin Robertson, one of the resident partners of the Company, regarding a few bird-skins he left for me at York. Certainly much obliged, having never seen him nor had any correspondence. Wrote him a note of thanks. Proceeded at eleven and passed some very bad rapids, and launched the boat at Trout Fall portage where the remainder of the day was spent repairing the boat. Laid in a few plants.

Friday 24th On leaving Trout Fall we found the boat still made water, but as the wind was favourable for passing through Knee Lake no time was lost. Went prosperously on and breakfasted on Knee near 'Tea Islands' so named from *Ledum palustre* being abundant to them. The scenery of this, like Oxford Lake, is fine also, with numerous islands.

Saturday 25th In the course of the morning the boat was considerably injured descending this stream, the water being low. Passed through Swampy lake, a pond a few miles long and then entered 'Hill River'. At noon, while descending a rapid, the boat struck heavily on the rock and shattered seven of the timbers and planking. Just had time to reach a small island when she was filled. My hands tied up — could not get off. Dried papers; planted in a small box *Erythronium grandiforum, Lilium pudicum,* and *Claytonia lanceolate,* which I am glad are all fresh. (Why did you not bring *Gaultheria* alive — across the continent — 2900 miles? It could be done.)

Sunday 26th Employed all day at the boat until three o'clock, when we set out again, the boat making a great deal of water. Camped a few miles below a low hill on the right.

Monday and Tuesday 27th and 28th Detained longer than usual, the morning being dull and unfavourable for passing the 'Rock Fall'. which was passed in safety at seven. Entered 'Steel River', a stream of some magnitude but not so rapid as the last. Breakfasted at its junction with York River. Continued until dusk, when we put to shore, boiled the kettle, and embarked under sail. Aurora borealis beautiful. The idea of finishing my journey, and expectations of hearing from England made the night pass swiftly. At sunrise on Tuesday I had the pleasing scene of beholding York Factory two miles distant, the sun glittering on the roofs of the house (being covered with tin) and in the bay riding at anchor the company's ship from England. The hearty welcome I had to the shores of the Atlantic from Mr McTavish and all others was to me not a little gratifying. In the most polite manner everything that could add to my comfort was instantly handed; and I adduce no further proof of this gentleman's goodness than to state that he had, without my knowledge, made for me a new suit of clothing, linen, &c. ready to put on.

No letters from England. Regret the death of my Calumet Eagle; was strangled a few days ago with the cord by which he was tied by the leg; fell over the casing of one of the houses and was found dead in the morning. What can give one more pain? This animal I carried 2000 miles and now lost him, I might say, at home. Had a note from Governor Simpson. Met Captain Black, Lieutenant Kendall, and Mr Drummond, who arrived yesterday. It now only remains to state that I have had great assistance, civility, and friendly attentions from the various persons I have formed an acquaintance with during my stay in North America.

8

Fame and the Final Years

Douglas returned to London to find that he was famous. Honours were heaped upon him and scientific London lionised him. For a time he was happy and he started organising his collection and editing his Journal which John Murray undertook to publish.

Unfortunately he slowly became rather dissatisfied, both with the Horticultural Society and his fellow scientists. He was underpaid and he began to see that the task of preparing the Journal for printing was beyond him. Similarly he made little progress with his collection, and among the unpublished Douglas papers in the Royal Horticultural Society's archives there is an account dated 16th April, 1829 recorded by Mr Goode on the very day of a tremendous row between Dr Lindley of the Society and Douglas. Hooker was called in to mediate and it is clear that Douglas was disciplined by his old sponsor.

William Hooker wrote a memoir of Douglas in 1836 which was published in the Companion to the Botanical Magazine, and he pulled no punches.

> "Qualified, as Mr Douglas undoubtedly was, for a traveller, and happy as he unquestionably found himself in surveying the wonders of nature in its grandest scale, in conciliating the friendship (a faculty he eminently possessed) of the untutored Indians, and in collecting the productions of the new countries he explored; it was quite otherwise with him during his stay in his native land. It was, no doubt, gratifying to be welcomed by his former acquaintances, after so perilous yet so successful a journey, and to be flattered and caressed by new ones; and this was perhaps the amount of his pleasures, which were succeeded by many, and, to his sensitive mind, grievous disappointments. Mr Booth remarks, in his letter to me on this subject, 'I may here observe, that his appearance one morning in the autumn of 1827, at the Horticultural Society's Garden, Turnham Green was hailed by no one with more delight than myself, who chanced to be among the first to welcome him on his arrival, as I was among the last to bid him adieu on his departure. His company was now courted, and unfortunately for his peace of mind he could not withstand the temptations

(so natural to the human heart) of appearing as one of the
Lions among the learned and scientific men in London; to
many of whom he was introduced by his friend and patron,
Mr Sabine. Flattered by their attention, and by the notoriety
of his botanical discoveries, which were exhibited at the meet-
ings of the Horticultural Society, or published in the leading
periodicals of the day, he seemed for a time as if he had
attained the summit of his ambition. But alas; when the
novelty of his situation had subsided, he began to perceive that
he had been pursuing a shadow instead of a reality.' As some
further compensation for his meritous services, the Council
of the Horticultural Society agreed to grant him the profits
which might accrue from the publication of the Journal of
his Travels, in the preparation of which for the press, he was
offered the assistance of Mr Sabine and Dr Lindley: and Mr
Murray of Albemarle-street was consulted on the subject. But
this proffered kindness was rejected by Mr Douglas, and he
had thoughts of preparing the Journal entirely himself. He
was, however, but little suited for the undertaking, and
accordingly, although he laboured at it during the time he
remained in England, we regret to say, he never completed
it. His temper became more sensitive than ever, and himself
restless and dissatisfied; so that his best friends could not but
wish, as he himself did, that he were again occupied in the
honourable task of exploring North-west America."

After Hooker's visit Douglas seems to have steadied, and
probably because of Hooker's influence, the Horticultural
Society decided to send him out on his travels again, with assistance from the Colonial Office and the Hudson's Bay Company.

The expedition this time was to range right up the Pacific
Coast from California northwards with particular reference to
the fauna of inland California.

Douglas was overjoyed, as his letters to Hooker (Appendix
I) show. Before he sailed he got expert instruction in surveying
from Sabine's brother Edward and learned the use of a wide
range of instruments.

Douglas may have been extremely tough and fit, but he
was having increasing trouble with his eyes and he found small
type very difficult. This may have been one of the causes of
his irritability — we shall never know. Suffice it to say that
a superficial study of his handwriting in the R.H.S. archives
indicates that it was steadily deteriorating with age.

He finally sailed for the Columbia in the Company's ship
Eagle on 31st October, 1829, accompanied by a small Scottish
terrier called Billy, calling at Honolulu on the way. Here he
was royally entertained and managed some trips ashore.

On 3rd June, 1830 he arrived back at the Columbia and

went up to Fort Vancouver. There had been many changes. Trees had been felled and land reclaimed and the Indian population greatly reduced by fever. It was never to recover. (See his letter of October 11th, 1830 to Hooker in Appendix I.)

We know that Douglas was still keeping a Journal, but this was lost in the accident on June 13th, 1833 in the Fort George Canyon as his letter to Hooker of May 6th, 1834 tells.

Indeed nearly all the knowledge of his activities, whereabouts and discoveries between 1830 and 1834 has to be gleaned from his letters and the records of his contemporaries.

Anyway, between June and November, 1830 he based himself on the Columbia and went around his old haunts — up the Snake River and into the Blue Mountains, up the Willamette and Santiam Rivers and down to Fort George. During this period he procured seeds of *Abies procera* — the Noble Fir — one of the most beautiful of all conifers. He had found it in 1825 in the Dalles of the Columbia, but collected seeds in the Cascade Mountains in the autumn of 1830, as he wrote to Hooker "I however transmit one bundle of six species, exceedingly beautiful of the genus *Pinus*. Among these, *A. nobilis* is by far the finest. I spent three weeks in a forest composed of this tree, and day by day could not cease to admire it; in fact, my words can be only monotonous expressions of this feeling".

Late in November he left Fort Vancouver on the Company's ship *Dryad* and reached Monterey in central California on December 22nd. He made a base for himself in the house of a Mr Hartnell, an English merchant, and in the next 18 months travelled the coastal region from Santa Barbara in the south up to Fort Ross, about 100 miles north of San Francisco. He made a number of highly important discoveries which have found their way into our gardens. He also named and collected *Pinus radiata, Pinus sabiniana, Abies venusta* and visited the great redwood forest *(Sequoia sempervirens)*.

His eyesight continued to deteriorate, but Douglas seems to have loved California and got on very well with the people he met. Harvey remarks that he arrived in California as Don Douglas, but left it as Dr Douglas because he skilfully set the arm of a little boy who had fallen into the hold of a ship in Monterey harbour. Certainly both in Hawaiian and in Californian literature Douglas is frequently referred to as Doctor.

There was a considerable sea trade between the Sandwich Islands and the North Western coast, so when the Company's ship did not arrive at Monterey he sailed for Honolulu.

On his arrival he learned that there had been considerable trouble in the Horticultural Society in London and that Sabine had been forced to resign. In sympathy Douglas also resigned. He was now a freelance, and he made a number of botanising trips which convinced him of the richness of the Islands. Nevertheless he only stayed for three weeks, and then took ship back to the Columbia.

Big changes had taken place at Fort Vancouver. Americans had started to come in and settle. A large ship was being built and farming and horticulture greatly extended. A school was being established for the half-breed children. Dr McLoughlin's policy of tolerance towards the American settlers was to be the cause of bitter rows between him and Governor Simpson, and in the long term it meant that all the huge territory now comprising the state of Washington became part of the United States.

In February Douglas penetrated up to Fort Nisqually at the southern end of Puget Sound and returned to Fort Vancouver. His eyesight was giving him more trouble, but he decided to have one last attempt and see if he could realise his ambition of walking back to Europe across Siberia and Russia (see letter to Hooker of August 6th, 1829 in Appendix I). Early overtures to the Russians had been favourable and Douglas intended to travel up the Columbia and the internal river systems of New Caledonia to the Russian settlement of Sitka in Alaska.

On March 20th he set off, but the trip turned out to be a series of accidents. His route went via Okanagan, to Kamloops. At Kamloops he had a violent quarrel with Chief Trader Samuel Black, but he pressed on across to Fort Alexander on the Fraser River. By June 6th he had reached Fort St James at the south end of Stuart Lake, and was made welcome by Chief Factor Dease. Here, however, a chance of further progress ended. He was still hundreds of miles from Fort Simpson, the nearest Company's post on the coast, and then he had a further three hundred miles to cover before he got to Sitka. Much of the country was unknown and the Indians were hostile. Regretfully he turned back. At Fort George he stopped for a couple of days and then set off by canoe down the Fraser. On June 13th disaster struck when his canoe was shattered in Fort George Canyon and he lost everything, (see Appendix I) and was more than fortunate to survive with his life. He set off again from Fort George a few days later and slowly retraced his steps the thousand odd miles to his base on the Columbia "greatly broken down, having suffered no ordinary toil". It did not stop him however making a further trip into the Blue Mountains, and, probably in late July, making the first recorded attempt to ascend Mount Hood.

On Friday 18th October, 1833 Douglas set sail from the Columbia in the Company's brig *Dryad* en route for the Sandwich Islands, never to return.

After calling at San Francisco, he reached Honolulu on December 23rd and spent Christmas with the British Consul Richard Charlton. By the New Year he was on Hawaii itself and basing himself at Hilo where he stayed with the American missionary Rev. Joseph Goodrich, he made plans to explore the interior and climb the great mountains of Mouna Kea (13784'), Mouna Loa (13680') and the active volcano of Kilaeua. In fact he climbed all three mountains within a month. Douglas kept a Journal of his activities in the Sandwich Islands which was published in Hooker's memoir. It is too long to reprint here, but I have included in Appendix I short excerpts from letters sent by him to Dr Hooker recounting some of his Hawaiian experiences.

It is hardly surprising that Douglas' health broke down after these exertions — his eyes in particular were now proving a great handicap and he returned to Honolulu to seek a passage for England.

There were no prospects of a ship for some weeks and Douglas resolved to show the Rev. John Diell, an American chaplain who served the Seamen's Mission, some of the island's wonders. A family party sailed from Honolulu on July 3rd, 1834 in the schooner Minerva. The Diells left the ship to visit the Island of Molokai having arranged a rendezvous with Douglas at Hilo on Hawaii. Douglas got put ashore at Kohala on the north tip of the island intending to cross over the mountain Mouna Kea (which he had previously climbed) to Hilo. He was accompanied by John, the Diell's negro servant, and his little dog Billy who had been with him ever since he left England.

John could not keep up with Douglas and they separated. After staying the night of July 11th at Nedo ranch Douglas moved on early and called and breakfasted at the hut of a certain Ned Gurney, an ex-Botany Bay convict who was making a living trapping wild cattle. Gurney urged him to wait for guides, but he was too impatient and so Gurney went a mile or so along the trail and warned him about the bullock pits he was bound to meet further on. One, which lay directly on the path was most carefully concealed, but two others nearby were open and each held a bullock. They parted.

Later in the morning passing natives were attracted to a torn piece of cloth above the concealed pit and looked in and saw the trapped bullock standing on the foot and shoulder of the body of Douglas. The earth had caved in and he had been gored to death. They called Gurney who shot the animal and extricated the corpse. Gurney hired natives to take the body to the sea and accompanied them. He then sent it on to Hilo by canoe where it arrived on July 15th. Goodrich and Diell were appalled at the injuries and their narrative letter to Richard Charlton on the 15th and 16th of July sets out their initial misgivings and their eventual conviction of Ned Gurney's innocence. The body was eviscerated however and sent to Honolulu for post mortem. It arrived on August 3rd badly decomposed. Nevertheless four doctors inspected it and agreed that the wounds had been inflicted by the bullock. On August 4th Douglas' body was interred in the burial ground of the native church. The funeral was attended by the foreign residents of Honolulu and the service read by one of the officers of H.M.S. Challenger.

There will always be a lingering doubt in some people's minds as to what really happened. I recently unearthed some papers in the library of the Royal Botanic Gardens in Edinburgh which suggested that Douglas was having an affair with Ned Gurney's wife, based on native hearsay, but there is not a shred of real evidence to support this. The possibility of murder was investigated and rejected within a month of his death. Suicide, particularly such a horrible suicide, is hardly credible.

In my view Douglas heard a trapped bullock in the pit, left his bag in the care of his little dog Billy a few yards back and went to investigate. Curiosity and short sight brought him too close to the edge, the side crumbled away, he slipped in and was gored to death.

Appendix I

EXCERPTS FROM DAVID DOUGLAS' LETTERS TO
DR. WILLIAM HOOKER 1829 - 1833

London, August 6th,
1829

I am sure you will be glad to know that my anticipated journey has been laid before the Council, and approved of; so that I go, God willing, on the 15th of September, by the Hudson Bay Company's Ship Eagle. My plans must be a separate communication, but just let me say, that my principal objects are to make known the vegetable treasures of the Interior of California, from the northern boundaries of Mexico, near the head of the Gulf. The botanical productions of Rio Colorado and other streams, totally unknown in Europe, will, I trust, ere many years be as familiar as those of the Columbia. The Government provides me with every instrument which Captain Sabine, as Secretary of the Royal Society thinks may be of use. These consist of sextants, chronometers, barometers, thermometers, hygrometers, compasses of all sorts, instruments for magnetic intensity, dip of the magnetic needle, all of which can be used with such accuracy, as will render my journey, as I trust, not the journey of a common-place tourist.

I am not quite certain, but that when I have completed my expedition on the Continent of America, I may cross to the opposite shore, and return in a southerly line, near the Russian frontier with China. What a glorious prospect! Thus not only the plants, but a series of observations may be produced, the work of the same individual on both Continents, with the same instruments, under similar circumstances, and in corresponding latitudes! I hope I do not indulge my hopes too far. I shall try to set a hundred pairs of feet, and as many of hands to work for me, and shall make them grub up and bring me all they can find. People tell me that Siberia is like a rat-trap, which there is no difficulty in entering, but from which it is not so easy to find egress. I mean at least to put this saying to the test. And I hope that those who know me know also that trifles will not stop me. I am glad to learn you are coming to England, before I go, that I may see you once more. I shall be greatly obliged if you would purchase for me a Bible, in 2 vols. 8vo., with a good bold legible type, and notes of reference, or more properly speaking marginal notes. I cannot see to read small type, and have been unable to find such an one in London, but I know there is a Scotch edition of the kind which I describe . . .

Greenwich, September 14, 1829

I am exceedingly engaged in my preparations, and will soon be ready. The vessel is to sail not later than the end of this month, which delights me amazingly. I go under most comfortable circumstances, and am certainly very happy. All my instruments are ready, save the chronometer, which I hope to be in possession of within a few days, all packed and ready to be sent on board ship at an hour's notice. Nothing pleases me so much as the addition of £20 which has been given me by the Colonial Office; I asked for £60 to provide books, tables, and charts, and they sent £80, as also some instruments, which though previously used by other persons, are in perfectly good order. I ought to think myself a very lucky fellow, for indeed every person seems to take more interest than another in assisting me. I possess a beautiful assortment of barometers, so constructed that, comparatively speaking, there is scarcely any liability of derangement, an object of most desirable attainment in these instruments. I shall combine observations accurately made with the hygrometer on different altitudes on the mountains and in different latitudes, which will, I trust, furnish you with information that can be confidently relied on, and which will effect much in illustrating the Geography of Plants . . . Did you hear of the total wreck of the Hudson Bay Company's ship on the sandbar at the entrance of the Columbia River, with the loss of every individual on board, forty-six in number, on the 11th of March last? It was the vessel in which Dr Scouler and I went out in 1824 when the late Captain was First-Mate. It is stated that those who escaped from the wreck were destroyed by my old friends, the Chenooks. This may be true, though I confess I entertain some doubts, for I have lived among those people unmolested for weeks and months. The temptation, however, of obtaining the wreck, may have overcome their better (if indeed they possess any) feelings. Though this is far from agreeable news and though the name of my new Captain (Grave) may sound ominous, I shall yet venture among these tribes once again. I doubt not I can do as much as most people, and perhaps more than some who make a parade about it. I shall write every day and write everything, so that my drivelling will return home, though perhaps I may not.

Entrance of the River Columbia
October 11th, 1830

How much I feel indebted to you for your long and kind letter of Christmas-day, 1829! I received it two months ago, four days after I had left headquarters for an extensive journey in the Cordilleras of New Albion, and what a stimulus it was to me! Situated as I am, without any one of kindred feelings, to share my labours and my toils and anxiety, such a letter makes all one's troubles seem light . . . I have now just saved the sailing of the ship, and, after sixty days of severe fatigue, have undergone, as I can assure you, one of still more trying labour, in packing up three chests of seeds, and writing to Mr Sabine and his brother. The Captain only waits for this letter, after which the ship bears away for Old England; I am truly sorry to see her go without my dried plants, but this is unavoidable, as I have not a bit of well-seasoned wood in which to place them, and should, moreover, be unwilling to risk the whole collection in one vessel; and the sails are already unfurled, so that it would be impossible to attempt dividing them. I however transmit one bundle of six species, exceedingly beautiful, of the genus *Pinus*. Among these, *A. nobilis* is by far the finest. I spent three weeks in a forest composed of this tree, and day by day could not cease to admire it; in fact, my words can be only monotonous expressions of this feeling. I have added one new species during this journey *A. grandis,* a noble tree, akin to A. *balsamea,* growing from one hundred and seventy to two hundred feet high. In the collection of seeds, I have sent an amazing quantity of all the kinds. Your specimens are in every way perfect. . . . The southern termination of the map is the source of the river, and the spot where, in October 1826, I had such a narrow escape from the hostile tribes, who inhabit that country. Since that time, a party of hunters were all killed, save two, who returned to tell the melancholy fate of their companions; and again a second party has nearly shared the same fate. You may judge of my situation, when I say to you that my rifle is in my hand day and night; it lies by my side under my blanket when I sleep, and my faithful little Scotch terrier, the companion of all my journeys, takes his place at my feet. To be obliged thus to accoutre myself, is truly terrible. However, I fail not to do my best, and if unsuccessful in my operations, can make my mind easy with the reflection that I used my utmost endeavours. My instruments are all excellent, and in the best order, and have already enabled me to make a multitude of important observations, which will go some way towards perfecting the Physical Geography of this part of the country, as well as illustrating its magnetic phenomena.

In Zoology, I possess some valuable additions to the Fauna, consisting of quadrupeds, birds, reptiles, and insects, which as well as the plants, must remain with me till next year.

A dreadful fatal intermittent fever broke out in the lower parts of the river about eleven weeks ago, which has depopulated the country. Villages, which had afforded from one to two hundred effective warriors, are totally gone; not a soul remains! The houses are empty, and flocks of famished dogs are howling about, while the dead bodies lie strewed in every direction on the sands of the river. I am one of the very few persons among the Hudson Bay Company's people who have stood it, and sometimes I think, even I have got a great shake, and can hardly consider myself out of danger, as the weather is yet very hot . . .

Farewell. I am daily, in recollection, with you and your family, though so unfortunate as to be divided from you by half the diameter of the globe; still the thought of you affords me, in my lonely walks, an inexhaustible source of delight.

I thank Mr Murray and Dr Scouler for their kind letters; to both I mean to write in spring, and shall send some articles of Comparative Anatomy to the latter.

Monterey, Upper California
November 23rd, 1831

. . . My whole collection of this year in California, may amount to five hundred species, a little more or less. This is vexatiously small, I am aware; but when I inform you that the season for botanizing does not last longer than three months, your surprise will cease. Such is the rapidity with which spring advances, as on the table-lands of Mexico and the platforms of the Andes in Ghili, the plants bloom here only for a day. The intense heats set in about June, when every bit of herbage is dried to a cinder. The facilities for travelling are not great, whereby much time is lost; this, as a matter of course, is the case in all new countries. It would require at least three years to do anything like justice to the Botany of California, and the expense is not the least of the drawbacks.

At present, it is out of my power to effect anything further, and must content myself with particularizing the collection now made. Of new genera I am certain there are nineteen or twenty, at least, and I hope you will find many more. Most of them are highly curious. As to species, about three hundred and forty may be new. I have added a most interesting species to the genus *Pinus, P. sabinii,* one which I had first discovered in 1826, and lost, together with the rough notes, in crossing a rapid stream on my return northwards. When compared with many individuals of the genus inhabiting the western parts of this continent, its size is inconsiderable, from 110 to 140 feet high, and 3 to 12 feet in diameter. In the aqueous deposits on the western flanks of the Cordilleras of New Albion, at a very great elevation above the sea (1600 feet below the line of perpetual snow), this Pine grows somewhat larger than in the more temperate parts near the coast in a more southern parallel. I sent to London a detailed account of this most beautiful tree, to be published in the Transactions of the Horticultural Society, which you will see before this can reach you, so that I will not trouble you with a further description of it. But the great beauty of Californian vegetation is a species of *Taxodium,* which gives the mountains a most peculiar, I was almost going to say awful, appearance — something which plainly tells that we are not in Europe. I have never seen the *Taxodium nootkatensis* of Nees, except some specimens in the Lambertian Herbarium, and have no work to refer to; but from recollection I should say, that the present species is distinct from it. I have repeatedly measured specimens of this tree 270 feet long and 32 feet round at three feet above the ground. Some few I saw, upwards of 300 feet high; but none in which the thickness was greater than those I have instanced. I possess fine specimens and seeds also. . . . As I shall, if it please God, have the happiness of writing to you again shortly, I will, at present, only tell you of my projects. I am daily in expectation of a vessel from Columbia, in which I shall embark to renew my labours in the North. Should she not arrive before the 10th of December, I will take my passage in an American vessel for the Sandwich Islands, where I shall not fail to endeavour to scale the lofty peaks of Mouna Roa or Mouna Kea (the White or Snowy Mountain) in quest of Flora's treasures, and proceed to the North West coast in the ensuing spring. I have met the Russian Authorities twice since I last wrote to you, and have received the utmost kindness from them. Two days ago I received a letter from Baron Wrangel, Governor of the Possessions in America and the Aleutian Isles, full of complaints, and offering me all manner of assistance, backed by Imperial favour from the Court. This nobleman is, as you are well aware, the Captain Parry of Russia, keenly alive to the interests of Science, and anxious to assist, in every way, those who labour in this field.

Since I began this letter, Dr Coulter, from the Central States of the Republic of Mexico, has arrived here, with the intention of taking all he can find to De Candolle at Geneva. He is a man eminently calculated to work, full of zeal, very amiable and I hope may do much good to Science. As a salmon-fisher he is superior even to Walter Campbell, of Islay, Esq., the Izaak Walton of Scotland; besides being a beautiful shot with a rifle, nearly as successful as myself! And I do assure you from my heart, it is a terrible pleasure to me thus to meet a really good man and one with whom I can talk of plants.

River Columbia
October 24th, 1832

. . .Mr Garry is exceedingly kind to me: I have also received a long letter from Capt. Sabine, dated Charlemont Fort, Ireland, full of kindness. Nothing can be more gratifying to me than to be remembered by old friends after the lapse of so many months, and when so far apart. Capt. Sabine goes so far as to say, that he can suggest to me no improvement in the manner of taking my astronomical or other observations, or in the way of recording them. He has shown them to the excellent Capt. Beaufort, who also expressed his approbation of them, and has (I fear, too partially) done the same officially to Mr Hay at the Colonial Office. Capt Sabine feels, I am sensible, too true a regard for my welfare not to point out my faults, and as this letter adverts to none, I may take it for granted, I trust, that he is well pleased with me. I have received a copy of Capt. Beechey's book. I entertain a great respect for that gentleman, but I think he has been too severe on the Catholic Missionaries in California. Any man who can make himself well understood by them, either in Castilian or Latin will discover very shortly that they are people who know something more than their mass-book, and who practise many benevolent acts, which are not a little to their credit, and ought to soften the judgment of the stranger, who has probably had more opportunity of seeing men and things than the poor priests of California. Their errors are the errors of their profession, and not of their hearts, and I thus make bold to say so, having had reason to know that the individuals in question are honourable exceptions to priests in general. I am no friend to Catholicism, still I should desire to maintain my own opinion without hurting the feelings of others . . . Not having received any letter from England, I cannot definitively state what will be the direction of my future journey. Should I receive no fresh orders, I shall, as I stated before leaving home, proceed to the northward of the Columbia, skirting the western flanks of the Rocky Mountains, as far as convenience and safety will allow, and endeavour to reach the sea to the westward, to some of the Russian Establishments or return by the same route, as may appear most desirable. On this point I shall be able to inform you in my next.

I have had two most kind letters from Baron Wrangel, Governor of the Russian Territories in America, and of the Aleutian Islands, to whom I was made known through the Russian Ministers at the Court of London. In the first he writes thus, which I know it will be pleasant to you to know, as it is highly agreeable to me:— "J'ai appris avec une vive joie votre intention de faire une tournée dans nos environs. Soyez sur, Monsieur, que jamais visite ne m'a été plus agréable, et que des bras ouverts vous attendent à Sitka. Si vous avez l'intention de retourner en Europe, par la Sibérie, je puis vous assurer qu'au mois de Mai de l'année prochaine, vous pourrez commodement aller sur un de nos navires à Okotsk, où, d'après des nouvelles que je viens d'apprendre, on vous a déjà preparé un gracieux accueil". This is more than kind, and the facilities offered for May, 1832, of course hold good for ensuing years. This letter was accompanied by a copy of a volume published in 1829, Recueil des Actes de l'Academie de St. Petersbourg, containing some very interesting accounts of the Russian expeditions to Mount Ararat; also an outline of Mertens' labours with Capt. Lutke's Pendulum and

Experiments made during his voyage. The Baron wrote me a sceond letter, and being fearful that I might not have received his first, took care to give me the same information backed with additional assurance of his good will. I have had the advantage of seeing Cyrill Klebinkoff, Chief Director of the Russian-American Fur Company, an excellent man, who has great claims on my gratitude, as well as several Officers of the Imperial Navy. Indeed they seem to be a set of people whose whole aim is to make you happy. You have my best thanks for replying to Dr Fischer of St. Petersburgh; I shall write to him when opportunity offers . . .

Interior of the River Columbia
April 9th, 1833

. . . Since I wrote to you, the season being winter, I have little new to communicate: during the interval I have made a journey, as I proposed, North of the Columbia, to New Georgia, and a most laborious one it was. My object was to determine the position of the Head Lands on the coast, and the culminating points of the many prodigiously high snowy peaks of the Interior, their altitudes, &c., and as I was favoured with exceedingly fine clear weather, this was effected much to my satisfaction. On this excursion I secured about two hundred species of Mosses; but as I am rather ignorant of this tribe, there may be a few more or less: certain it is, however, that there are many fine kinds that are totally unknown to me; and perhaps even you may find some of them new. I have also some interesting Fuci from Puget's Sound, collected on the same journey, three of which are decidedly not in Mr Turner's work, and very noble species they are. I have bespoken the services of all the Captains on the North West coast, to bring me all sorts of sea-weeds, simply coiled up, dried, and put in a bag . . . I proceed to give you a short sketch of my intended movements this year.

As soon as the season permits, which I trust will be in a few days, I shall leave this spot for the northward, travelling sometimes in canoes, or on horse back, but far more generally on foot. The country is mountainous and very rugged, the rivers numerous and there are not a few lakes of considerable extent. Perhaps I shall cross Mackenzie's track, at Fraser's River (called the Columbia by that great traveller), in about long 122° West and proceed northwards, among the mountains, as far as I can do so with safety, and with the prospect of effecting a return. The country is certainly frightful; nothing but prodigious mountains to be seen: not a deer comes, say the Indians, save once in a hundred years — the poor natives subsist on a few roots. My outfit is five pounds of tea, and the same quantity of coffee, twenty-five pounds of sugar, fifteen pounds of rice, and fifty pounds of biscuit: a gallon of wine, ten pounds of powder and as much of balls, a little shot, a small silk fishing net, and some angling tackle, a tent, two blankets, two cotton and two flannel shirts, a handkerchief, vest, coat, and a pair of deer-skin trousers (not those kindly presented to me by Dr Gillies, which, by repeated exposure to the rain shrunk so much that I was reluctantly obliged to give them away), two pairs of shoes, one of stockings, twelve pairs of mocassins and a straw hat. These constitute the whole of my personal effects; also a ream and a half of paper, and instruments of various kinds; my faithful servants, several Indians, ten or twelve horses, and my old terrier, a most faithful, and now to judge from his long grey beard, venerable friend, who has guarded me throughout all my journies, and whom, should I live to return, I mean certainly to pension off, on four pennyworth of cat's-meat per day!

I am most anxious that you should know what I see and do on this important journey, and as it may so turn out that I shall never have the pleasure of meeting you more, I intend, God willing, to commence writing a little to you on the very first evening of my journey, which is fixed for the 18th, and continue thus to condense, from time to time, the substance of my notes, putting down whatever may appear most important and interesting to me.

Fever still clings to the native tribes with great obstinacy, and not a few of the people of the Hudson's Bay Company have suffered very severely from it. Only three individuals out of one hundred and forty altogether escaped it, and I was one of that smaller number. Thank God, I never was in better health, and could I but have a few moments with you, I might add, in excellent spirits. Even the employment of writing to you, tends to enliven my mind. It is singular, that while my left eye is become infinitely more delicate and clear in its power of vision, the sight in my right eye is utterly gone; and, under every circumstance, it is to me as dark as midnight. If I look through a telescope or microscope, I generally see objects pretty well at a short distance, but the least fatigue brings on a doubling of the object with a surrounding vapory haze, that soon conceals everything. These results were owing to an attack of ophthalmia, in 1826, followed by snow-blindness, and rendered irretrievable by the scorching heat of California. I use purple goggles to diminish the glare of the snow, though most reluctantly, as every object plants and all, is thus rendered of the same colour . . .

                                        Woahoo, Sandwich Islands
                                        May 6th 1834

. . . You will probably enquire why I did not address you by the despatch of the ship to Europe last year. I reached the sea-coast greatly broken down, having suffered no ordinary toil, and, on my arrival, was soon prostrated by fever. My last letter to you was written from the interior of the Columbia, and bore date about the middle of April, 1833 (last year), just before starting on my northern journey. Therein I mentioned my intention of writing a few lines to you daily, which I did, up to the 13th of June, a most disastrous day for me, on which I lost, what I may call, my all! On that morning, at the Stony Islands of Fraser's River (the Columbia of McKenzie — see the map in his 4to. edition), my canoe was dashed to atoms, when I lost every article in my possession, saving an astronomical journal, book of rough notes, charts, and barometrical observations, with my instruments. My botanical notes are gone, and what gives me most concern, my journal of occurrences also, so this is what can never be replaced, even by myself. All the articles needful for pursuing my journal were destroyed, so that my voyage for this season was frustrated. I cannot detail to you the labour and anxiety this occasioned me, both in body and mind, to say nothing of the hardships and sufferings I endured. Still I reflect, with pleasure, that no lives were sacrificed. I passed over the cataract and gained the shore in a whirlpool below, not however by swimming, for I was rendered helpless, and the waves washed me on the rocks. The collection of plants consisted of about four hundred species — two hundred and fifty of these were mosses, and a few of them new. This disastrous occurrence has much broken my strength and spirits. The country over which I passed was all mountainous, but most so towards the Western Ocean: — still it will, ere long, be inhabited . . .

I sailed from the Columbia in November last, in the Hudson Bay Company's vessel, which visited these islands, touching on the way at San Francisco, where I made a short stay, but did nothing in the way of Botany. I arrived here on the 23rd of December, and, after spending Christmas Day with two English ladies, the wife of our Consul, Mr Charlton, and her sister, I started on the 27th for the island of Hawaii, which I reached on the 2nd of January, 1834. You know I have long had this tour in contemplation, and having spent three winter months in botanising here, I proceed to give you a short notice of my proceedings.

The view of this most interesting island, from the sea, is sublime indeed, combining the grand, sweet, and beautiful, in a most remarkable degree. For two thousand five hundred feet above the level of the sea the Banana, Sugar Cane, Coffee, Pandanus, Bread Fruit, &c. grow in the greatest perfection. Then comes a thickly timbered country as high as eight thousand feet, and for three thousand seven hundred feet more, a space covered with short verdure; after which the reign of Flora terminates. I made a journey to the summit of Mouna Kea, which occupied fourteen days, and found it only thirteen thousand eight hundred and fifty-one English feet above the sea; a height, you may observe, much less than has been ascribed to this mountain by early travellers. In this expedition I emassed a most splendid collection of plants, principally Ferns and Mosses: many, I do assure you, truly beautiful, and worthy to range with the gigantic species collected by Dr Wallich. Of Ferns alone I have two hundred species, and half as many Mosses; of other plants comparatively few, as the season is not yet good for them, nor will be so, until after the rains. On my return, I must consult with you on the best mode of publishing the plants of these islands.

I also visited the summit of Mouna Loa, the Dig or Long Mountain, which afforded me inexpressible delight. This mountain, with an elevation of thirteen thousand five hundred and seventeen feet, is one of the most interesting in the world. I am ignorant whether the learned and venerable Menzies ascended it or not, but I think he must have done so, and the natives assert that this was the case. The Red-faced Man, who cut off the limbs of men, and gathered grass, is still known here; and the people say that he climbed Mouna Loa. No one, however, has since done so, until I went up a short while ago. The journey took me seventeen days. On the summit of this extraordinary mountain is a volcano, nearly twenty-four miles in circumference, and at present in terrific activity. You must not confound this with the one situated on the flanks of Mouna Loa, and spoken of by the Missionaries and Lord Byron, and which I visited also. It is difficult to attempt describing such an immense place. The spectator is lost in terror and admiration at beholding an enormous sunken pit (for it differs from all our notions of volcanos, as possessing cone-shaped summits, with terminal openings), five miles square of which is a lake of liquid fire, in a state of ebullition, sometimes tranquil, at other times rolling its blazing waves with furious agitation, and casting them upwards in columns from thirty to one hundred and seventy feet high. In places, the hardened lava assumes the form of gothic arches in a colossal building, piled one above another in terrific magnificence, through and among which the fiery fluid forces its way in a current that proceeds three miles and a quarter per hour, or loses itself in fathomless chasms at the bottom of the cauldron. This volcano is one thousand two hundred and seventy-two feet deep; I mean down to the surface of the fire; its chasms and caverns can never be measured. Mouna Loa appears, indeed, more like an elevated Table-land than a mountain. It is a high broad dome, formed by an infinitude of layers of volcanic matter, thrown out from the many mouths of its craters. Vegetation does not exist higher than eleven thousand feet; there is no soil whatever, and no water. The lava is so porous that when the snow melts, it disappears a few feet from the verge, the ground drinking it up like a sponge. On the higher parts some species of *Rubus, Fraseria, Vaccinium,* and some *Junci.*

I visited also the volcano of Kiraueu, the lateral volcano of Mouna Loa; it is nearly nine miles round, one thousand one hundred and fifty-seven feet deep, and is likewise in a terrific state of activity.

I go immediately to Hawaii to work on these mountains. May God grant me a safe return to England. I cannot but indulge the pleasing hope of being soon able, in person, to thank you for the signal kindness you have ever shown me. And really were it only for the letters you have bestowed on me during my voyage, you should have a thousand thanks from me . . .

Appendix II

BIOGRAPHICAL NOTES ON DOUGLAS' ACQUAINTANCES

William Conolly 1787(?) - 1849

Born Lachine, near Montreal. Joined N.W. Co. 1801, partner 1818. 1821 made Chief Trader; 1825 Chief Factor in charge of New Caledonia. In 1805 married a Cree woman. Six children, one of whom married James Douglas. 1832 Conolly repudiated his first wife and married his cousin. A famous law case ensued some years later.

Sir James Douglas 1803 - 1877

Born Demerara, British Guiana. Educated at Lanark. In 1818 joined North West Company and in 1821, Hudson Bay Company. Associated John McLoughlin. Founded Victoria on Vancouver Island 1843. Governor Vancouver Island 1851. 1858 first Governor of British Columbia. Married William Conolly's half-breed daughter. 1863 K.C.B.

Edward Ermatinger 1797 - 1876

Fur Trader. Author. Born Elba, of Swiss/English descent. Educated in England, he and his brother Francis (1798 - 1758) joined the Hudson's Bay Company in 1818. Settled in Upper Canada in 1830. Wrote a "Life of Col. Talbot" and his son, Judge C. O. Ermatinger, edited and published his father's York Factory Express Journal in 1912.

Jaco Findlay ???? - 1828

Douglas met Findlay at Spokane in 1826. He was the half-breed son of the James Findlay who explored the Saskatchewan. Jaco was with Thomson when he penetrated the Howse Pass in 1807. He built Spokane House about 1810 and died there in 1828.

Sir John Franklin 1786 - 1847

The celebrated Arctic explorer whom Douglas met and travelled with in 1827. He led numerous expeditions to the North West from 1818 onwards, finally perishing near King William Land in his 62nd year. By all accounts a delightful companion — in the words of the Dictionary of National Biography "a man not only of iron resolution and indomitable courage, but of a singular geniality, uprightness and simplicity".

William Jackson Hooker 1785 - 1865

One of the truly great 19th century English scientists, and father of Joseph Dalton Hooker, the friend and associate of Charles Darwin. A man of great generosity and enthusiasm and posessed of quite exceptional energy, he is chiefly remembered today for the virtual creation of the Royal Botanic Gardens at Kew. Travelled extensively in Europe as a young man. In 1820 he was appointed to the Chair of Botany in

Glasgow where his lectures became famous. In Glasgow he first noticed David Douglas, a young gardener on the staff of the Botanic Gardens. He quickly recognised Douglas' qualities and showed him much personal kindness. He acted as a patron to the youth and secured for him his first job as a plant collector for the Horticultural Society. He kept up a regular correspondence with Douglas for the rest of his life, and helped him through various difficulties. Author of the first account of Douglas' life in 1836. He wrote a number of highly important botanic works. Knighted 1836. His son took over the management of Kew on his death. For a full account of this wonderful man read Mea Allen's *The Hookers of Kew.*

J. G. Mactavish b. (?), d. 1847

A great character and bosom friend of Governor Simpson. Scot. Joined North West Company 1798. Helped in the rescue of Thomson in 1813. Secured the surrender of J. J. Astor's trading post Astoria in 1813. In charge of York Factory 1821-1828, where Douglas met him. Lived with two native women by whom he had many children before he married a Scots girl in 1830.

Finan Macdonald 1782 - 1851

Fur trader. Born Aberdeenshire. Joined North West Company. Associated with David Thompson on the Columbia 1807 - 1812. Joined Hudson's Bay Company in 1821 and travelled the Columbia until 1827, when he settled in Upper Canada. Married an Indian.

Donald McKenzie 1783 - 1851

Governor of Assiniboine. Scots born, he emigrated to Canada in 1800 and joined the Hudson's Bay Company. In 1809 he joined J. J. Astor's company and in 1811 went to Astoria. Joined the North West Company in 1813 when it took Astoria. In 1821 became a Chief Factor.

John McLoughlin M.D. 1784 - 1857

One of the great men of the region. Originally in the employment of the North West Company he became after the merger with the Hudson's Bay Co. Chief Factor for the Columbia River Department. Founded Fort Vancouver in 1825, but later on fell out with Governor Simpson. An independent and liberally minded man, he was greatly respected by the Indians (he was called by them the White Headed Eagle) and later by the first American settlers whom he helped with loans. A Catholic, and married to a half-caste, he was a giant of a man, tough, irascible and soft hearted. Simpson's account of meeting him on the Express route in 1823 is worth quoting:

"He was such a figure as I should not like to meet on a dark night in one of the bye lanes in the neighbourhood of London, dressed in clothes that had once been fashionable but now covered with a thousand patches of different colours . . . loaded with arms and to his own herculean dimensions forming a tout ensemble that would convey a good idea of the highway men of former days".

He later described him as a somewhat disorganised manager, scrupulously honest, but that he had an ungovernable violent temper and "would be a radical in every country under any Government and under any circumstances".

His memory is rightly revered and respected in Oregon to this day. Sometimes called the "Father of the Oregon".

John Norbet Provencher 1787 - 1847

Douglas met Provencher on the Red River in 1827. French Canadian priest, who founded the College of St. Boniface near Winnipeg in 1820. In 1820 appointed Roman Catholic bishop of the North West.

Sir John Richardson 1787 - 1865

Explorer and naturalist. Born in Dumfries. M.D. Edinburgh 1817. Accompanied Sir John Franklin on his first two Arctic explorations 1819 - 22, 1825 - 27. In 1848 commanded the early part of the Franklin Search Expedition. Wrote the Natural History Notes of this expedition.

Edward Sabine 1788 - 1883

Younger brother of Joseph Sabine, he was a distinguished scientist in his own right and was particularly interested in the measurement of the earth's magnetic field. F.R.S. in 1818, sailed with Ross in 1818 and Parry in 1819. Secretary of Royal Society 1827. Taught Douglas the use of survey instruments in 1829. Later became a general and was knighted. Wrote many scientific papers. Sabine's Gull is named after him.

Joseph Sabine 1770-1837

A lawyer by profession, but drawn to science by inclination and during the early part of the 19th century a most important figure in the scientific world. One of the original members of the Linnean Society. F.R.S. In 1810 took over the chaotic affairs of the Horticultural Society and reorganised it most efficiently as Honorary Secretary. An autocrat, by 1830 he had over-reached himself and heavy debts had been incurred. He resigned, but played a significant part in the Zoological Society until his death. A useful patron and friend of David Douglas who wrote many letters to him from his travels. Sabine wrote several papers and his name is commemorated in a number of plants and trees.

Sir George Simpson 1787(?) - 1870

Born in the parish of Loch Broom in Wester Ross. Illegitimate. Brought up by relations near Dingwall and went to London as a youth. Took over the Athabasca District of the Hudson's Bay Company in 1820 in a time of great trouble with the North West Company. After the merger he rapidly rose to a position of power. His remarkable journey along the Hudson Bay express route in 1824 must rate as a classic. Leaving York Factory in mid August, he reached the Pacific coast at Fort George (Astoria) on November 8th, knocking twenty days off the previous fastest passage. Immensely tough, hard and unscrupulous. Douglas met him in 1827 when he was Governor of the Northern Department, and already very powerful. Formed many liaisons with Indian women before marrying in 1830, in London, his cousin. As his position became un-assailable he became more and more involved in politics and industry. Knighted in 1841, he died in Montreal in 1860.

John Stuart 1779 - 1847

Born in Strathspey, he joined the North West Company in 1799 and in 1806 accompanied Simon Fraser on his epic exploration of the Fraser River. Stuart Lake is named after him. He returned to Scotland in 1839 and died at Springfield House near Forres in 1847. Met Douglas in 1827 en route for York Factory. A great pioneer. A Chief Factor. 1825 - 1833 Governor of Assiniboine.

John Work 1792 - 1861

Irish born. Joined Hudson's Bay Company in 1814. In 1823 went to the Columbia and spent the rest of his life west of the Rockies. Married a Spokane half-breed. 1846 made Chief Factor. Elected to executive council in Vancouver in 1857.

Appendix III

PLANTS INTRODUCED TO BRITAIN BY DAVID DOUGLAS

I am indebted to Dr. Ian Hedge, the Keeper of the Herbarium of the Royal Botanic Gardens in Edinburgh and his staff for checking earlier lists. As far as possible, the names of the North West American species have been brought into line with the nomenclature adopted in Munz, P. A., *A Californian Flora* and Hitchcock, C. L. et al., *Vascular plants of the Pacific North West* vols. 1 - 5 (1955 - 1969).

LATIN NAME	POPULAR NAME
Abies amabilis	Lovely fir
grandis	White fir
procera	Noble fir
Abronia mellifera	Sand verbena
Acer circinatum	Vine maple
macrophyllum	Broad-leaved maple
Agastache urticifolia	Giant hyssop
Amelanchier alnifolia	Juneberry (Saskatoon)
Amsinckia lycopsoides	Fiddle-neck
Anemone multifida	Globose anemone (Wind Flower)
Antirrhinum multiflorum	Sticky snapdragon
Arbutus menziesii	Madrona
Arctostaphylos columbiana	Manzanita
Astralagus crassicarpus	Rattle weed
Berberis aquifolium	Tall Mahonia (Tall Oregon
Berberis nervosa	Grape)
Biosduvalia densiflora	
Brassavola nodosa	A dwarf orchid
Brodiaea congesta	Wild hyacinth
coronaria	(four species)
hyacinthina	
laxa	
Callandrinia ciliata var. menziesii	Redmaids
Calochortus albus	Mariposa lily or tulip
luteus	(six species)
macrocarpus	
pulchellus	
splendens	
venustus	
Camassia quamash	Camas
Castilleja parviflora	Small-flowered paint-brush
Clarkia amoena	Clarkia (Farewell to Spring,
lepida	Summer's Darling)
pulchella	(six species)
purpurea ssp quadrivulnera	
purpurea ssp virninea	
rhomboidea	

LATIN NAME	COMMON NAME
Clematis virginiana	Virginia clematis
Collinsia hetrophylla	Chinese houses
grandiflora	Blue-eyed Mary
parviflora	
Collomia grandiflora	Collomia
linearis	
Coreopsis atkinsoniana	Atkinson's coreopsis
Cornus stolonifera	Red-osier dogwood
Crataegus douglasii	Hawthorn
Delphinium menziesii	Menzies larkspur
Dendromecon rigida	Bush poppy
Douglasia nivalis	Mountain pink
Downingia elegans	Downingia
pulchella	
Epilobium minutum	Willow herb
Erigeron speciosus	Showy fleabane
Erogionum compositum nudum	Sulphur flower
Eriophyllum lanatum	Woolly sunflower
Erythronium grandiflorum	Dog-tooth violet
Eschscholzia caespitosa	Californian poppy
californica	
californica var. douglasii	
Fritillaria pudica	Mission bell
Gaillardia aristata	Gaillardia
Garrya elliptica	Quinine or tassel brush
Gaultheria shallon	Salal
Geranium carolinianum	Carolina geranium
Gesneria douglasii	
Gilia archilleifolia	Gilia (10 species)
ssp. multicaulis	
aggregata	
androsacea	
capitata	
coronopifolia	
densiflora	
liniflora pharnaceoides	
tenuiflora	
tricolor	
Gomesa planifolia	An orchid
Harkelia congesta	
Helianthus annuus	Sunflower
Heuchera cylindrica	Alum root
micrantha	
Holodiscus discolor	Ocean spray
Horkelia congesta	
Hosackia crassifolia	Pink bird-foot clover

LATIN NAME	COMMON NAME
Iris tenax	Tough-leaved iris
Lasthenia glabrata	Lasthenia
Lathyrus japonicus	Beach pea
Layia chrysanthemoides	Tidy tips
Lespedeza capitata	Bush clover
Limnanthes douglasii	Meadow foam
Linum perenne	Perennial blue flax
Lonicera ciliosa	Orange honeysuckle
hirsuta	Hairy honeysuckle
hispidula	Purple honeysuckle
Lupinus benthanii	Lupin (18 species)
bicolor	
chamissonis	
densiflorus	
hirsutissimus	
latifolius	
laxiflorus	
lepidus	
leucophyllus	
littoralis	
micranthus	
nanus	
polyphyllus var. albiflorus	
polyphyllus var. polyphyllus	
rivularis	
sabinii	
sericeus	
sulphureus	
Lotus pinnatus	
Madia elegans	Madia
Malus fissea	Pacific coast crab apple
Meconella linearis	Platystigma
Mentzelia lindleyi	Blazing star
Mikania scandens	Climbing hempweed
Mimulus cardinalis	Mimulus or monkey-flower
floribundus	(five species)
guttatus	
lewisii	
moschattus	
Nemophila menziesii	Baby blue eyes
Nicotiana multivalvis	Native tobacco
Oenothera albicaulis	Evening primrose (five species)
biennis var. muricata	
dentata	
speciosa	
triloba	
Oncidium pubes	A tropical epiphytic orchid

LATIN NAME	COMMON NAME
Paeonia brownii	Wild peony
Pedicularis canadensis	Lousewort
Penstemon acuminatus	Beard-tongue or penstemon
attenuatus	(19 species)
breviflorus	
centranthifolius	
confertus	
deustus	
fruticosus var. fruticosus	
fruticosus var. scouleri	
glandulosus	
gracilis	
heterophyllus	
nemorosus	
ovatus	
pruinosus	
richardsonii	
serrulatus	
speciosus	
triphyllus	
venustus	
Phacelia divartica	Heliotrope (four species)
linearis	
tanacetifolia	
viscida	
Phlox longifolia	Phlox
Picea sitchensis	Sitka spruce
Pinus contorta	Lodgepole pine
coulteri	Big cone pine
lambertiana	Sugar pine
monticola	Western white pine
ponderosa	Western yellow pine
radiata	Monterey pine
sabiniana	Digger pine
Platystemon californicus	Creamcups
Pogonia pendula	An orchid
Potentilla arguta	Cinquefoil (five species)
hippiana	
glandulosa	
pensylvanica	
pensylvanica var. strigosa	
Prunus pumila	Sand cherry
Pseudotsuga menziesii	Douglas fir
Psoralea macrostachya	Leather root
orbicularis	
Purshia tridentala	Antelope bush
Ribes aureum	Golden currant
cereum	Squaw currant
divaricatum	Straggly gooseberry
divaricatum var. irriguum	Black gooseberry

LATIN NAME	COMMON NAME
gracile	Pasture gooseberry
lacustre	Swamp currant
malvaceum	Chaparral currant
menziesii	Prickly currant, swamp gooseberry
petiolare	Western black currant
sanguineum	Flowering Red Currant
sanguineum var. glutinosum	Winter currant
setosum	Red-shoot gooseberry
speciosum	Fuchsia-flowered gooseberry
viscosissimum	Sticky currant
Rubus leucodermis	Black raspberry
parviflorus	Thimbleberry
spectabilis	Salmonberry
Salvia dorrii var. camosa	Gray-ball sage
Sidalcea malviflora	Marsh hollyhock
Sinningia helleri	A gloxinia
Spergula arvensis	Common spurrey
Sphaeralcea munroana	False mallow
Stylomecon heterophylla	Flaming poppy
Symphocarpus albus	Snowberry
Tanacetum douglasii	Huron tansy
Trifolium fucatum	Sour clover
Vaccinium ovatum	Evergreen huckleberry
Xerophyllum tenax	Bear grass

Appendix IV

Among the Douglas papers in the archives of the Royal Horticultural Society are some notes that Douglas made in 1828 and 1829 on a few of the North West American conifers. These notes were printed in full in the 1914 edition as Appendix VIII. Douglas called all conifers pines and sometimes confused the species with known European or Eastern American trees. Systematic botany has developed a great deal during the last hundred and fifty years and during that period some trees have had their generic and specific names changed several times. I have used the latest nomenclature.

The following notes are much abbreviated but are of some importance to botanists and foresters.

1. *Pseudotsuga menziesii* (Douglas fir)

Tree remarkably tall, unusually straight, having the pyramid form peculiar to the *Abies* tribe of Pines. The trees which are interspersed in groups or standing solitary in dry upland, thin, gravelly soils or on rocky situations, are thickly clad to the very ground with widespreading pendent branches, and from the gigantic size which they attain in such places and from the compact habit uniformly preserved they form one of the most striking and truly graceful objects in Nature. Those on the other hand which are in the dense gloomy forests, two-thirds of which are composed of this species, are more than usually straight, the trunks being destitute of branches to the height of 100 to 140 feet, being in many places so close together that they naturally prune themselves, and in the almost impenetrable parts where they stand at an average distance of five square feet, they frequently attain a greater height and do not exceed even 18 inches in diameter close to the ground. In such places some arrive at a magnitude exceeded by few if any trees in the world generally 20 or 30 feet apart. The actual measurement of the largest was of the following dimensions: entire length 227 feet, 48 feet in circumference 3 feet above the ground, 7½ feet in circumference 159 feet from the ground.

Some few even exceed that girth, but such trees do not carry their proportionate thickness to such a vast height as that above mentioned. Behind Fort George, near the confluence of the Columbia River, the old establishment of the Honourable the Hudson's Bay Company, there stands a stamp of this species which measures in circumference 48 feet, 3 feet above the ground, without its bark. The tree was burned down to give place to a more useful vegetable, namely potatoes.

Being an inhabitant of a country nearly in the same parallel of latitude with Great Britain, where the winter (even judging from the immense covering afforded it by Nature in its bark) is more severe, gives every reason to hope that it is in every respect well calculated to endure our climate and that it will prove a beautiful acquisition to English Sylva if not an important addition to the number of useful timbers.

The wood may be found very useful for a variety of domestic purposes: the young slender ones exceedingly well adapted for making ladders and scaffold poles, not being liable to cast; the larger timber for more important purposes; while at the same time the rosin may be found deserving attention.

2. *Picea sitchensis* (Sitka spruce)

The appearance of this species closely resembles *P. menziesii,* although neither so large nor so plentiful as that species, it may nevertheless become of equal if not greater importance. It possesses one great advantage by growing to a very large size on the Northern declivities of the mountain in apparently poor, thin, damp soils; and even in rocky places, where there is scarcely a sufficiency of earth to cover the horizontal wide-spreading roots, their growth is so far from being retarded that they exceed one hundred feet high and eight feet in circumference. This unquestionably has great claims on our consideration as it would thrive in such places in Britain where even *P. sylvestris* finds no shelter. It would become a useful and large tree.

3. *Abies procera* (Noble fir)

The present tree, among the many highly interesting species by which it is surrounded in its native woods, in point of elegance justly claims the pre-eminence. The trees are straight, one hundred and seventy feet high, two to six feet in diameter with a white smooth polished bark. An open-growing tree, sparingly clad with wide-spreading horizontal branches placed in regular whorls round the tree, the distance between the whorls diminishing towards the top. The cones are always on the highest branches, near the top like *P. Picea* and *P. balsamea.*

The wood is soft, white and very light, containing but little rosin.

An inhabitant only of the mountains, seldom if ever seen to arrive at any considerable size lower down on the hills than 5000 feet above the level of the sea in the 46 degrees and 48 degrees N. Lat.

Common on a chain of mountains that run nearly parallel with the coast, from Arguilar River on the South to the base of Mount St. Helens on the North and in the intermediate distance intersecting Mount Hood and Mount Vancouver. Not on the Rocky Mountains? This grows more luxuriantly and more abundantly in thin poor dry soil on a rocky bottom; near springs or rivulets it entirely disappears.

This if introduced would profitably clothe the bleak barren hilly parts of Scotland, Ireland and Cumberland, besides increasing the beauty of the country.

4. *Abies amabilis* (Beautiful or Red fir)

This is another tree of singular beauty, not quite so tall but exceeding in girth the preceding one. This does not much exceed one hundred and thirty feet, but is sometimes seen eight feet in diameter, a very graceful and more compact tree than the former. The branches are very long, drooping and flat, the leaves maintaining the same character. Is well contrasted with the distinct and varied foliage of those that surround it. The bark of the full-grown timber is partly green with large blotches, smooth, producing little or no rosin. That of the young trees is smooth and polished green, with minute round or oval scattered blisters, yielding a limpid fluid, which possesses a less pungent taste and has a less aromatic odour than Balm of Gilead Fir *P. balsamea,* to which in some instances it is related. The wood is soft, white, very light, but heavier than the last mentioned, clean-grained, and takes a good polish, and under the microscope shows a multitude of minute angular reservoirs filled with the like fluid contained in the blisters but somewhat more viscid and less fragrant.

This inhabits the same range, and is equally as plentiful as the last-mentioned species, and though not quite so fine, justly merits our further consideration.

5. *Pinus monticola* (Western White pine)

A handsome tree of large dimensions, particularly in aqueous deposits consisting of decayed vegetable substances in the mountain valleys which are washed by the torrents from higher altitudes, also in rocky, bare, thin soils it particularly abounds. They grow immensely large, frequently 5 feet in diameter, and a hundred and sixty feet high, and, as is the case with many others, they naturally prune themselves, leaving a clean trunk, without a single branch, exceeding one hundred feet.

A common tree in the mountainous districts of the Columbia, from its confluence with the sea in 46° 10' N. Lat. to its source in 52° 30'; also on the banks of Flathead River, and the western base of the Rocky Mountains.

A truly distinct and beautiful species intermediate between the present and *Pinus lambertiana,* exists with cones as long as, if not longer than it, of an elegant slender taper form about 2 inches in diameter, it was only seen on one of my journeys to the mountains near the base of Mount St. Helens, I never had it in my power to procure perfect specimens of this desirable tree.

6. *Pinus ponderosa* (Western Yellow pine)

Trees tall, straight, seldom divided by large branches, very elegant, ninety to one hundred and thirty feet high; sometimes exceeding four feet in diameter, three feet above the ground, carrying their thickness to a very great height, frequently measuring eighteen inches in diameter at seventy feet, and not uncommon to see this without so much as a branch of any description whatever.

The wood is remarkably clean-grained, though somewhat coarse in texture, smooth, heavy, reddish, works fine, and is impregnated with a copious rosin.

The bark is very smooth, tawny-red. This appears to be rapid-growing, is highly ornamental, and may, though not so valuable as some, be of importance. Like all the species of this genus which have plural leaves, that inhabit the western parts of the continent of America, it never grows in nor composes thick forests like the *Abies* section, but is found on declivities of low hills and undulating grounds in unproductive sandy soils in clumps, belts, or forming open woods, and in low, fertile, moist soils totally disappears. This may have greater claims on our attention than merely its beauty, for, in addition to its timber, a great portion of turpentine could be extracted.

M

7. *Pinus contorta* (Shore pine)

Trunk 20 to 30 feet high, one foot to eighteen inches in diameter, with a rough, white bark, seldom divided by its branches. Wood soft, spongy, coarse-grained, brown-coloured, with abundance of rosin. Branches drooping, greatly twisted in every direction, remarkably tough, the younger ones covered by acuminate chaffy brown scales. As far as my opportunities of observing this species went, it is exclusively the inhabitant of the sea coast from 44° to the 49° of North Latitude, at all times confined to damp boggy soils where *Vaccinium uliginosum, Oxycoccus macrocarpus* delight to grow and where there is uniformly a dense carpet of *Sphagnum obtusifolium* and tufts of *Bartramia*. Not uncommon from "Cape Look Out" to the confluence of the Columbia; more abundant towards Puget Sound, and very likely will be found to increase in number to the 60°.

Little can be said in favour of this tree either for ornament or as a useful wood.

Index

A

C

Bibliography

The main work on Douglas' life is A. G. Harvey's scholarly *Douglas of the Fir* which embodies much loving research and yet is very readable. It was published by the Harvard University press in 1947 and gives an excellent bibliography which has been most useful.

Unlike Harvey I have had the privilege of examining the original Journals and material in the archives of the Royal Horticultural Society and the Linnean Society. Harvey drew his account largely from the 1914 comprehensive publication which had been sponsored by the Royal Horticultural Society. It is an indigestible book and I hope that one day a scholar will re-edit all the original material.

My main sources have been:

Allan, Mea, *The Hookers of Kew*. Michael Joseph. London. 1967.

Bubner, J. Bartlet, *Canada*. University of Michigan Press. Ann Arbor. 1960.

Desmond, R., *Dictionary of British and Irish Botanists and Horticulturalists*. Taylor and Francis. London. 1977.

Dictionary of Canadian Biography. 3rd edition. Macmillan. London and Toronto. 1963.

Douglas, David, Journal kept by David Douglas during his Travels in North America 1823 - 1827. Wesley & Son. London. 1914.

Dow, R., *David Douglas, Scone, Botanist & Pioneer of Arboriculture*. Transactions and Proceedings of the Perthshire Society of Natural Science 1909 - 1910.

Elwes J. J. and Henry A., *The Trees of Great Britain & Ireland*. 7 vols. Privately printed Edinburgh. 1906 - 1913.

Fraser, Esther, *The Canadian Rockies*. M. G. Hurtig. Edmonton. 1969.

Galbraith, S., *The Little Emperor*. MacMillan. Toronto. 1976.

Harvey, A. G. *Douglas of the Fir*. Harvard University Press. 1947.
    David Douglas in British Columbia. B.C.H.Q. Victoria B.C. 1940.

Hooker, W. J., *A Brief Memoir of the life of Mr David Douglas with Extracts from his letters*. Companion of the Botanical Magazine. London. 1836.

Holloway, D., *Lewis & Clark and the crossing of North America*. Pinnell Book Service. London. 1974.

Kerr, D. G. G. & Davidson, R. I. K., *Canada — a visual History*. Nelson. Toronto. 1966.

Mitchell, A. F., *Conifers in the British Isles*. H.M.S.O. London. 1972.

Websters American Biographies. G. & C. Merriam Company. Springfield Mass. 1974.

Wilson, W. F., *A Glimpse of Douglas at Monterey*. Compiled from David Douglas at Hawaii.

I have also read William Norwood's *Traveller in a Vanished Landscape*, Gentry Books, London, 1973, a somewhat imaginative account of Douglas' life, part of which at least is at variance with the known facts.